Butterfield Overland Mail Route

History of Early Settlers Along Boonville Road
in Northern Greene County

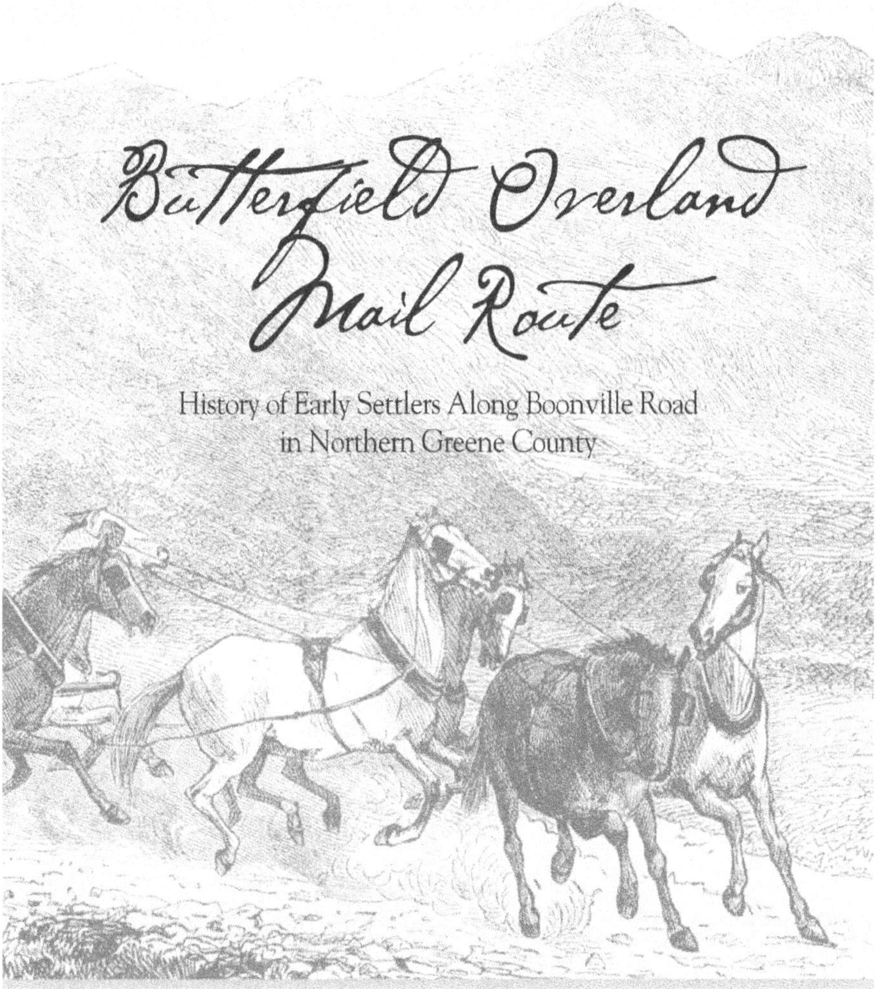

By Helen Murray White

HERITAGE BOOKS
2014

HERITAGE BOOKS
AN IMPRINT OF HERITAGE BOOKS, INC.

Books, CDs, and more—Worldwide

For our listing of thousands of titles see our website
at
www.HeritageBooks.com

Published 2014 by
HERITAGE BOOKS, INC.
Publishing Division
5810 Ruatan Street
Berwyn Heights, Md. 20740

Copyright © 2014 Helen Murray White

Editing and design by TheBookArtists.com
Maggie Castrey & Lean McKay

All rights reserved. No part of this book may be reproduced or transmitted in any form or by any means, electronic or mechanical, including photocopying, recording or by any information storage and retrieval system without written permission from the author, except for the inclusion of brief quotations in a review.

International Standard Book Numbers
Paperbound: 978-0-7884-5561-2
Clothbound: 978-0-7884-9018-7

Dedication

This book is dedicated to David and Huldah Murray and their many descendants.

Contents

Chapter 1. *All Roads Lead to Boonville*	1
Chapter 2. *Boonville Road and the Butterfield Mail*	20
Chapter 3. *Boonville Road Becomes Bolivar Road*	38
Chapter 4. *Early Settlers in Sections 35 and 26*	47
Chapter 5. *The Banfield Family*	78
Chapter 6. *David Murray Comes to Missouri*	89
Chapter 7. *Burials in the Murray Cemetery*	106
Epilogue	121
Index	125

Illustrations and Pictures

Map showing the Cumberland Road in 1825	6
Map in Past and Present of Greene County, Missouri, by Fairbanks and Tuck	7
1835 government survey Section 11, Township 29, Range 22	10
Survey report of Sections 35, 26, 23, Township 30, Range 22	11
Surveyor's Map showing Sections, 14, 11, 3	12
Missouri map of 1844 by Huttawa	14
Picket's Map of 1862	17
Butterfield Overland Mail Route through Missouri	22
David Murray Map 1867	23
Settlers in Section 14, Township 30, Range 22	24
Photos of Evans' Station	29
1998 Missouri Atlas and Gazetteer	30
Colton's Map 1869	32
Lloyd's Military Map 1861	33
Overland Mail Route sign at Dickerson Park Zoo	35
1876 Greene County Plat Book	41
Diagram of proposed road change 1904	43
Proposed road change at Little Dry Sac River in 1928	44
Proposed road change at Little Sac River in 1928	45
Plat of Section 35	48
Condolence Letter from Tapley Daniel to Huldah Murray	59
Plat of Section 26	63
Gravestones of Roger Q. Banfield & Lucy A. Banfield	85
Photo of David Murray	89
Receipt for papers and money due Office of Treasurer, September 1866	91
Photo of David Murray home in Ohio, 1985	93
Picture of Huldah Murray	96
Italianate farm house published in Canada Farmer Magazine, 1865	99
Photo of Huldah Murray's House circa 1900	99

Receipt in divorce settlement between Maggie Gay and William Gay	100
David Murray tombstone, Margaret Murray tombstone, Unnamed Infant	107
Bill for Margaret Murray's coffin	108
Freddie Murray tombstone	108
Order for tombstones from Buffalo Marble Works February 1890	109
Descriptions of tombstones ordered 1890	110
Bill for workers to clear the cemetery	111
Stone for Willhmina & Racie Brune	112
Louis Dickens tombstone	112
Butterfield Route on Murray Farm	121
Kiosk #1 at David C. Murray Trailhead on South Dry Sac Greenway	121
Kiosk#2 at David C. Murray Trailhead on South Dry Sac Greenway	122
Huldah Murray's House 2013	123

Preface

When I was a child, I spent hours going through the family trunk, examining discarded clothing, family pictures and bags of papers containing letters, tax receipts, promissory notes and bank books. They didn't make much sense, but they piqued my interest in learning about the unfamiliar names on the documents.

One of the most informative documents was a drawing of our farm, which must have been given to my great grandfather David Murray when he purchased the land in 1867. It showed Section 35, Township 30, Range 22, Greene County, Missouri, with two rivers, Little Sac and Dry Sac, converging at the west side of Section 35. Going through the property was a line labeled, "Boonville Road."

I grew up in the red brick farmhouse built by my great grandmother. The house sat adjacent to the road and my Dad said that the old Bolivar Road passed in front of the house. There were bridge pilings left from the first bridge to cross the Sac River, which was the bridge built prior to construction of old Highway 13, now Farm Road 141. Was the Bolivar Road the same as the Boonville Road? It had to be, there were no other highways matching the configuration found in David Murray's map.

So I began my search to learn about the Boonville Road and the Bolivar Road. I found that the Boonville Road was used by the Butterfield Overland Mail and passed in front of our old farmhouse.

This book is a search for the origins of the Boonville Road and to learn why the name changed to Bolivar Road. It traces the Butterfield Overland Mail Route as it traveled through northern Greene County. The Boonville Road crossed sections where the earliest settlers purchased property and married their neighbors. Their names are as familiar to me as if they were family. I also wanted to identify people who were buried in the Murray farm cemetery and I found that it was originally a neighborhood cemetery, not a family cemetery. Some of the people buried there were family members, some were neighbors, and some were buried without stones and could only be identified by newspaper obituaries. Others will always remain unknown.

Acknowledgements

This project has been a labor of love for many years. It all began when I found those letters and papers in the family trunk; I am indebted to the grandmothers who saved everything when papers could have so easily been destroyed.

Don Matt advanced the project one step further when he contacted me about researching the location of Evans' Station. His information on the Butterfield Overland Mail opened a new avenue of research on the Boonville Road.

Thanks to the dedicated employees who answered my questions and searched for documents: Robert Newman and staff at the Greene County Missouri Archives, John Rutherford and staff at the Genealogy Room, The Library Center, and James Coombs at the Missouri State University map library and Jeff Patrick, Wilson's Creek National Battlefield. Susan Sparks, Polk County Genealogical Society helped me look through United States government survey reports in Polk and Dade counties. J. Dale West conferred with me about his research in Lawrence County.

Thanks to the Greene County Missouri Historic Sites Board for placing Huldah's house and the lane in front of her house on the Historic Sites Register and the Springfield-Greene County Park Board for its foresight in creating a trailhead, which as well as being a significant recreational site, also focuses on the history of northern Greene County.

Special thanks to my husband, Bill, who endured hours sitting in courthouses, viewing genealogical records and listening to me read various re-writes. He provided support and encouragement when the research became more extensive than I had originally planned.

Helen Murray White

Chapter 1

All Roads Lead to Boonville

When the Louisiana Purchase was completed in 1803, the United States acquired land that included the future state of Missouri. People started for the West seeking land and opportunity and by the end of the War of 1812, settlers were pouring into the land along the Missouri River. According to Holcombe, "In 1810 a colony of Kentuckians numbering 150 families immigrated to Howard County and settled on the Missouri River near the present town of Franklin and opposite Arrow Rock." (1) One observer noted that it was as if the states of Kentucky and Tennessee were being emptied into Central Missouri. The Territory of Missouri began surveying land in 1816, and although some settlers were reluctant to purchase land on which they believed Native Americans still had claims, most were happy to purchase land as soon as it was made available.

Conflicts arose very quickly between the early settlers and the Indians. Although the Sauk and Fox were present in Northern Missouri, the dominant Indian tribe in the area south of the Missouri River and northern Arkansas was the Osage. As the settlers poured in, the Osage were forced into the Treaty of 1808, which essentially gave away all their lands in Missouri.

The treaty read: The chiefs and the warriors of the Great and Little Osage agree that the boundary line between the two nations shall be: "Beginning at Fort Clark, on the Missouri, five miles above Fire Prairie, and running thence a due south course to the river Arkansas, and down the same to the Mississippi, hereby ceding and relinquishing forever to the United States, all the lands which lie east of the said line, and north of the southwardly bank of the said river Arkansas, and all lands situated northwardly of the river Missouri." (2)

In return, the United States agreed to furnish a blacksmith, tools to mend their arms, utensils of husbandry, to build them a horse mill or water mill and furnish them with plows, and promised to protect the Osage from other Indian tribes.

The U.S. government promised to pay $1000 yearly in merchandise to the Great Osage and $500 yearly in merchandise to the Little Osage. The United States acknowledged that it had already paid $800 to the Great Osage and $400 to the Little Osage.

A treaty with the Osage did not solve the problems for the settlers living in central Missouri around the Missouri River and its tributaries. They still had to deal with the Sac, Fox, Kickapoo and Potawatomi Indians who harassed and killed the newcomers. For protection, the settlers built their homes and farms around forts so they could flee to the forts when attacked. Conditions changed by 1819 when immigrants started arriving in large numbers. According to the author, "They came in wagons, in carriages, in pirogues, and finally on every puffing steamer that ascended the turbid waters of the Missouri." (3) He quoted from the Franklin *Intelligencer* 19 November 1819, "It is calculated that the number of persons accompanying these wagons, etc., could not be less than 3,000." (4)

With increased population and help from the U.S. military moving the remaining Indians further west, the settlers were able to move away from their forts and establish independent communities. From the Fayette *Intelligencer* 2 May, 1828: "The town of Franklin, as also our village [Fayette], presents to the eye of the beholder, a busy, bustling and commercial scene, in buying, selling and packing goods, practising mules, etc., all preparatory to the starting of the great spring caravan to Santa Fe. A great number of our fellow citizens are getting ready to start, and will be off in the course of a week on a trading expedition." (5)

In preparation for trading to the west, citizens of the newly settled areas petitioned Congress for military posts for the "encouragement and protection between Missouri and the internal provinces of Mexico." Requests were made by the citizens of Boone County, 24 January 1825. On 25 January 1825, a similar request was made by Senator Benton on behalf of the citizens of Howard County. (6)

The U.S. Congress passed a law 3 March 1825, which was "to authorize the President of the United States to cause a road to be marked out from the Western frontier of Missouri to the confines of New Mexico." (7) No time was wasted in signing a new treaty with the Osage. It was signed by the chiefs of the Great Osage, Little Osage and U.S. Commissioners on 10 August 1825. The chiefs agreed that the citizens of the United States and the Mexican Republic should travel the road without interference by the Great and Little Osage. The road passed through the territory to which they had just been removed.

Prior to the treaty signing, in 1824, Colonel Matthew Arbuckle, Captain of the 7th Infantry, established a military post on the east side of the Grand River about a mile north of its confluence with the Arkansas River. This was in response to petitions by citizens requesting protection from Indians on the western border of Missouri. The post became known as Fort Gibson, Indian Territory, now Oklahoma, and was used to protect the border and to maintain peace between the Cherokee and Osage.

Because people were moving west, churches saw an opportunity to expand their congregations. In 1820, the United Foreign Missionary Society, an organization supported by the Presbyterian, Reformed Dutch and Associated Reformed churches, established a Mission School on the Neosho River, near Fort Gibson, Indian Territory, for the purpose of converting the Osage.

Then in 1821, the same organization established a mission near Pappinsville, Benton County, Missouri, known as Harmony Mission. The individuals who founded this mission traveled from Pennsylvania to Missouri by floating down the Ohio River, traveling up the Mississippi to the Missouri River, then down the Osage River, and on to the place where the mission was built. (8) The journey from Pittsburgh to Pappinsville took 112 days.

Harmony Mission school began with two Osage pupils and increased to fifty five pupils, but the mission was never successful. The Osage were not interested in planting crops and plowing fields. The men felt it was dishonorable for a man to do much besides hunt and go to war. Harmony Mission was finally discontinued in 1836.

It is obvious when reading early U.S. statutes that in the early years of the democracy the United States government felt it was necessary to build post roads and military roads. In reviewing the records of the Senate and House of Representatives, every session of Congress had requests by senators and representatives to provide post roads to their respective states. On 3 March 1825, the U.S. Congress established a post road from St. Louis to Boonville passing by the seat of justice in the county of Gasconade. (9)

For the early settlers, transportation occurred primarily by water, which made settlement so attractive along the Missouri River and its tributaries. In 1816, the little town of Franklin was laid out on the north side of the Missouri River, opposite the present site of Boonville. Franklin grew rapidly and for a time was the largest and most flourishing town in the state. It became the origin of the Santa Fe Trail.

A man named A. Fuller wrote his friend Tom, from Franklin, December 1819 and described conflicts between the settlers and Indians, but still believed that he would not change places with those who lived in luxury. He then described land sales. "They have laid out a town opposite here on the river, called Boonville, which they expect to eclipse this place, but the traders think Franklin will eclipse any town out west. I think likely it will if the river will let it alone. I went over the river last summer to attend the first sale of lots, but they were run up to a fabulous price." (10)

However, the river did not let Franklin alone. Floods and shifting waters gradually washed away the little town and commerce was transferred to the south side of the river, to the town named Boonville, which in turn took over as the starting point for the Santa Fe Trail and became a supply center for the entire Southwest Territory. W. F. Johnson described the waterfront, "A few years after 1826, the year in which the waters of the turbulent Missouri commenced encroaching upon the beautiful city of Franklin, Boonville assumed its dominant position on the Santa Fe Trail. Steamboats began to land in increasing numbers along the river front, especially at the foot of what is now Main Street, and there continued for years a wonderful activity."(11) Now it is obvious why the citizens of Howard County and Boone County wanted a military road to the southwest and protection for travelers. Boonville was the transportation hub, which explains why "Boonville Road" became so important and frequently mentioned. Early books and maps indicate that roads leading in the direction of Boonville were called Boonville Road, which makes research difficult and sometimes questionable.

In newly organized Greene County, at the second session of the County Court, 12 March 1833, a public road was ordered viewed and marked out from Springfield to the 25-mile prairie, in the direction of Boonville. The 25-mile prairie is located in Section 24, Township 35, Range 24, Polk County, Missouri. The previous day, the Court appointed commissioners to lay out a road from Bledsoe's ferry, on the Pomme de Terre River, to an indefinite point on the 25-mile prairie. Absolom Bledsoe was granted a license to keep a ferry across the Osage River and charge for every foot passenger 12 ½ cents, for every two-horse team $1.25. (12) This would indicate a road connection to Boonville.

This is further confirmed in the *History of Benton County*, when the author states that in 1831, Lewis Bledsoe, "located where the old road crossed the Osage, about one mile and a half above Warsaw and established a ferry." (13)

The same author says that the first settlers called the road the "old road" or "old military road." He states: "From these names my inference is that the road was originally cut out by the U.S. government for military purposes. It extended from Palmyra, on the Mississippi River, to Boonville, to Springfield, to Fayetteville, Arkansas and finally Fort Smith, Arkansas." (14)

There are no references to Palmyra in the U.S. statutes, as a beginning point for a military road. Palmyra was mentioned as a location on a post road in 1836. The reference to Palmyra may have presumed the town to be a part of the plan for completion of the Cumberland Road, or National Road connecting the East Coast to Missouri. However, it was never completed that far and Palmyra, located in Marion County, was too far north to be a part of the Cumberland Road.

Congress passed an appropriations bill 3 March 1825, for the building of the Cumberland Road to Zanesville, Ohio AND the extension of surveys to the permanent seat of government in Missouri. Missouri and Illinois fought over whether the crossing of the Mississippi River should be at Alton or St. Louis and, because the exact location could not be decided, the project was never completed. (15)

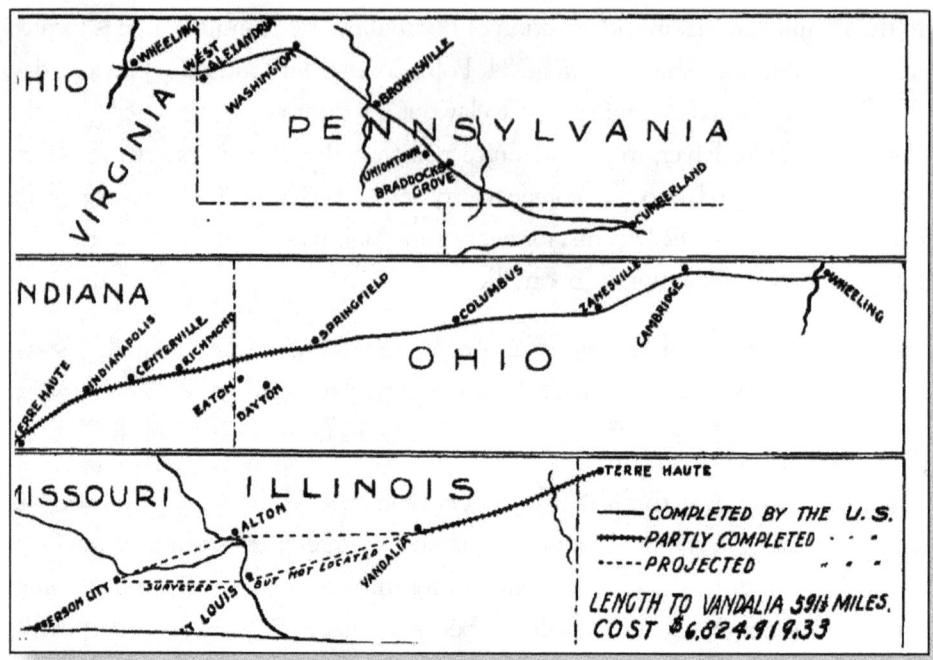

Map showing the Cumberland Road in 1825. The top two maps show the eastern section of the road. The bottom map shows the continuation from Terre Haute, Indiana, and the projected road to Jefferson City, Missouri.

In the text, *Past and Present of Greene County, Missouri*, by Fairbanks and Tuck, a map with data supplied by Edward M. Shepard and drawn by A.M. Haswell, shows a road leaving Springfield going north.

The road is labeled by various names: "Boonville Road, Osage Indian Trail, Old Military Road cut by govt in 1825."(16) The Justus Moll collection, located at the State Historical Society of Missouri-Columbia Research Center in Columbia, has a photocopy of the same map. (17) There is no documentation provided for the source of the information.

However the map published in Fairbanks and Tuck differs from what looks like the original map compiled by Dr. Shepard. This map is located at The Library Center, Springfield, Missouri. There is no date on the map, only the label, which states "Aboriginal data by Edw. M. Shepard, drawn by A. M. Haswell." That map states "road cut by govt in 1835." (18) Fairbanks and Tuck state that the road was "cut

Early Settlers Along Boonville Road

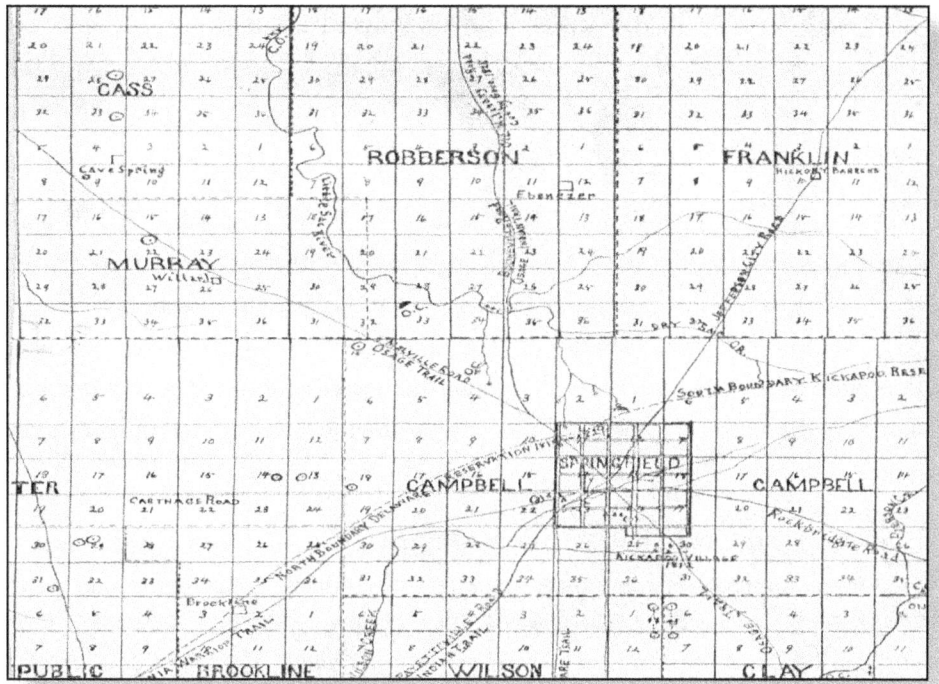

Map *in* Past and Present of Greene County, Missouri, *by Fairbanks and Tuck (16)*

out to the legal width by Act of March 7th, 1835." (19) It appears that the 1835 date is correct, and the 1825 date printed on the map in Fairbanks and Tuck was incorrect.

Fairbanks and Tuck repeat the route given by James H. Lay in his *History of Benton County*, writing that the road extended from Palmyra to Fort Smith and citing Lay even though he had offered no documentation. Lay **inferred** that because early settlers referred to the road as "old road" or "old military road," it must be the correct name.

The road description of the "Old Military Road cut by the government in 1825" seems to have been transferred from author to author, with no documentation provided. There were no US statutes passed in 1825 creating a military road within Missouri although a military road was established in that year running from the western border of Missouri to New Mexico.

It is more believable that the settlers followed Indian Trails, hacking out vegetation wide enough for their wagons to pass through the thick forests. They assigned the name of the road based on its final destination; hence there are many references to Boonville Road in Southwest Missouri.

In March 1836, Congress established a detailed list of post roads in Missouri. One of the roads ran from "Jefferson City, the seat of Government through the seats of justice for the counties of Morgan, Benton, Polk and Greene, to the seat of Government of the Territory of Arkansas." (20) This might have been our Boonville Road, but it is doubtful because there were several roads originating at towns on the Missouri River. References to Boonville Road in government survey reports and early Greene County records had been made prior to 1836.

In addition to the roads, frequent steamboats navigating the Osage River were also used for commerce. *The Saturday Morning Visitor* was a newspaper published in Warsaw, Missouri and is included in the Digital Newspaper Collection at the State Historical Society of Missouri-Columbia Research Center. The 10 June 1848 issue published the steamboat schedule for arrivals and departures: Arrivals included the *St. Louis Oak* and the *Wave*, both from the mouth of the Osage located on the Missouri River and the *Wave* from Osceola and Harmony Mission. Departures included the *St. Louis Oak* from Warsaw to St. Louis; the *Wave* from Osceola and Harmony Mission; and later, the *Wave* to the mouth of the Osage.

Mr. H.C. Henry, advertising in the same newspaper, said his goods were "cheap for cash and quick sales," and advised those going by wagon to Boonville to "give me a call and examine my goods thoroughly." Another advertisement stated: "Arrival of *Steamer Lightfoot*, carrying a large general assortment of staple groceries, in part—rectified whiskey, brandy, cognac, wine, candy, lemon syrup, coffee, sugar, shoes and boots." These advertisements demonstrated a diverse system of delivering goods to the Southwest with strong competition between overland wagons from Boonville and steamboats on the Osage River.

That the Boonville Road was widely used for commerce was also illustrated in a recollection found in *Ozar'Kin*, published by the Ozarks Genealogical Society. D.D. Berry was the leading merchant in the county in the 1830s and his store did a business of about $10,000 per annum. The retail prices were marked up 100 per cent over cost. All goods were received by river in Boonville and hauled in wagons to Springfield. (21)

In March 1834, the Greene County Court decided in a matter of a road leading from Springfield in the direction of Boonville. The order read: "On motion it is ordered by the Court here that the road leading from Springfield in the direction of Boonville be so altered or amended as to leave the present new road at the Junction of John Mooney and Jerimiah N. Sloans spring branches, running thence so as to leave Sloans field, then taking the dividing and running with same until it comes near William Slagles and there to intersect the old road the same is declared to be a public road or highway in Greene County." (22) The location is now in Polk County, but was part of Greene County in 1834, near the present small town of Slagle. This indicates a county highway, rather than a state highway, although later the Missouri legislature did determine some locations along this road. Mooney and Sloan were early settlers in the area even though their land patents were not issued until 1845-48.

In 1835 the U.S. government began surveying sections in Greene County. (23) In November 1835 the government surveyor surveyed Section 11, Township 29, Range 22. Section 11 is where Springfield is located. A clearly marked line reads "Road to Boonville." He surveys to the middle part of Section 3 and the line abruptly stops. Field notes in the surveyor's report mention "road bearing to Boonville" and describe prairie, timberland and quality of land for cultivation. There is no indication why the Boonville road is not shown continuing through Section 3 or why there are no further field notes after he stops drawing the Boonville road in Section 3. (See page 10)

In a field document dated 18 March 1835, the government surveyor recorded boundary lines in Sections 34, 35, 26, 27, Township 30, and Range 22. He noted a wagon road "leading a little west of south from Boonville to Springfield, the seat of justice of Greene County." A little further he sets a limestone as quarter section marker in Mr. Lastley's field. Section 35 is immediately north of Section 3.

The map on page 11 shows Boonville Road going through Sections 35, 26 and 23, the two rivers converging in Mr. Lastley's field in Section 35, which abuts Section 3, Township 29, Range 22, shown in the first map.

November 1835 government survey Section 11, Township 29, Range 22

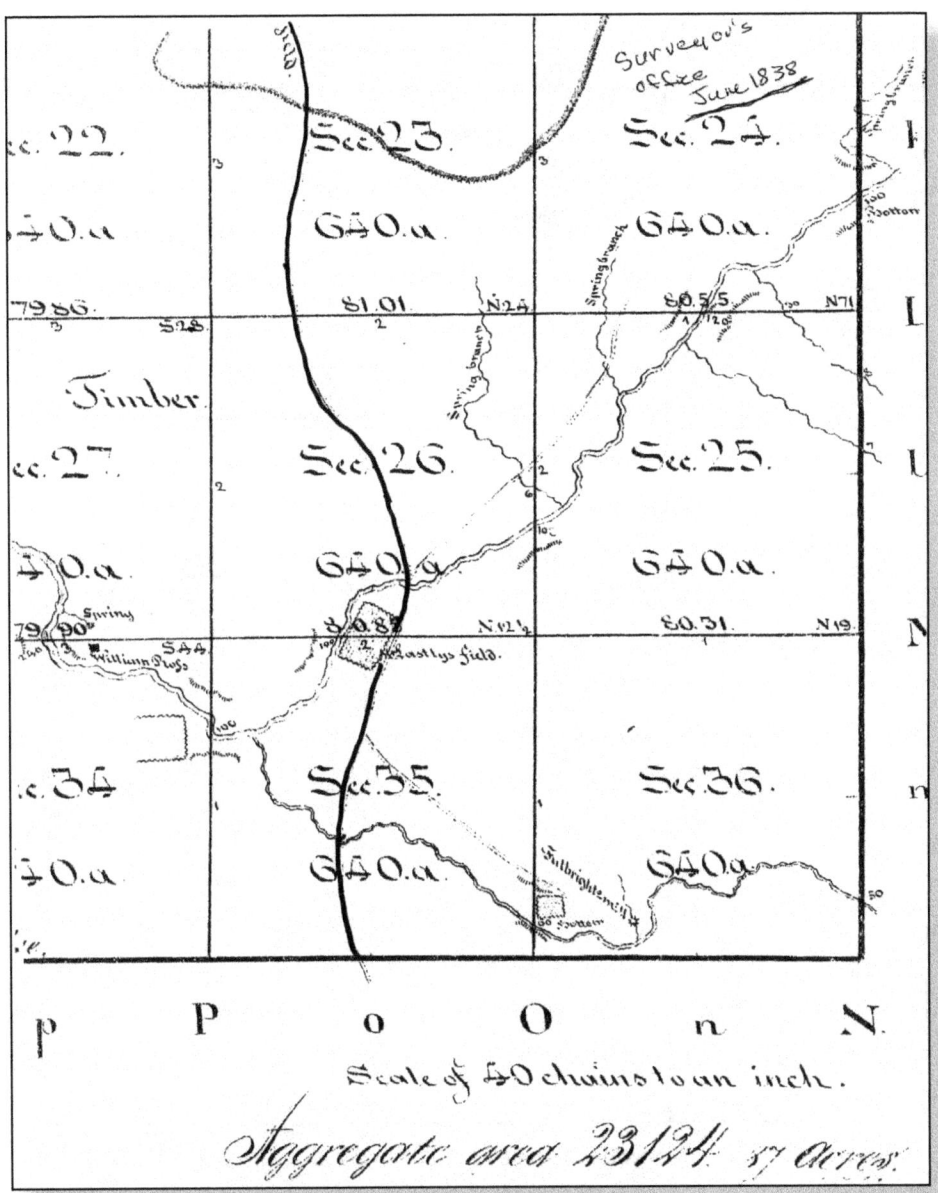

Survey report of Sections 35, 26, 23, Township 30, Range 22

The next map shows the sections continuing north to the Polk County line.

The surveyor's map shows "Road from Boonville to Springfield" passing through Sections 14, 11 and 3. Note the road going through the west half of Section 14.

Butterfield Overland Mail Route

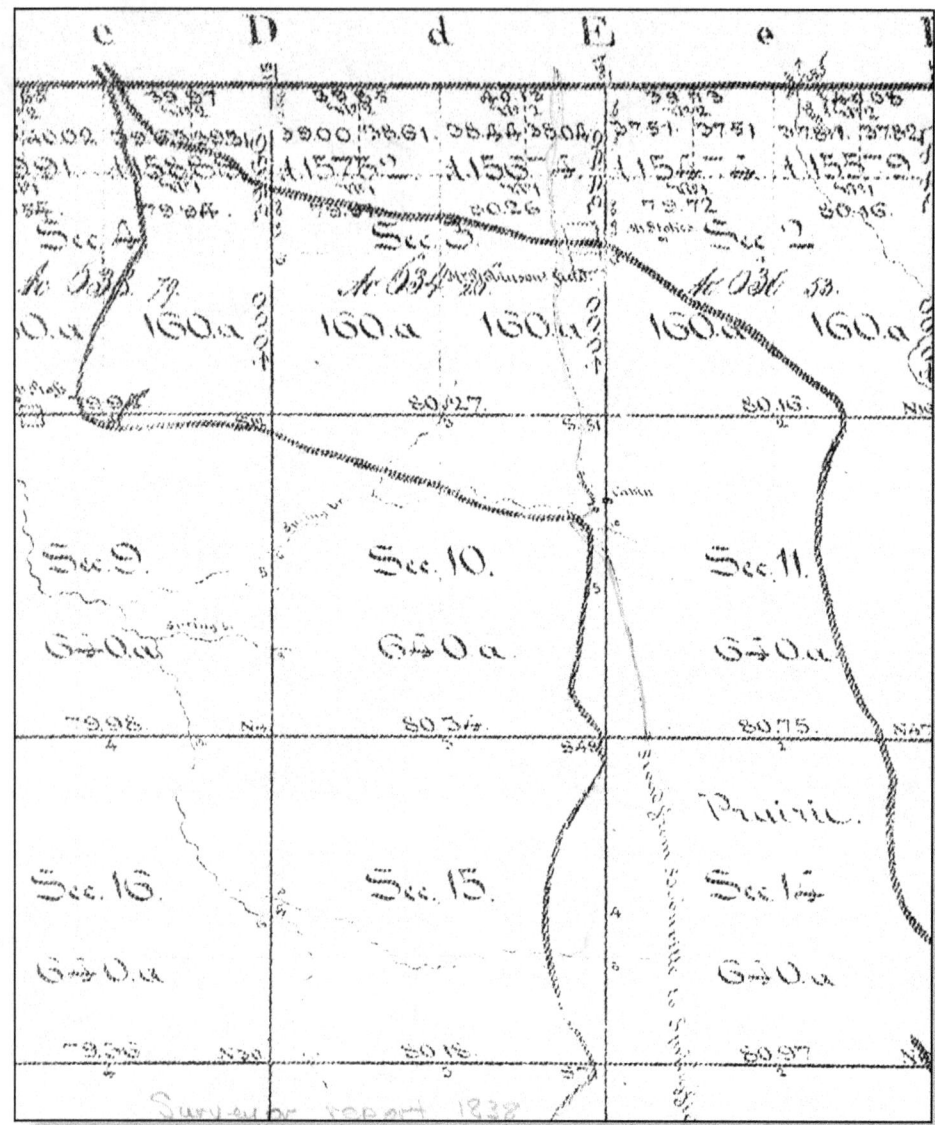

Surveyor's Map showing Sections 14, 11, 3. Section 14 abuts Section 23 on the previous map.

The northwest quarter of the southwest quarter is where Evans' Station on the Butterfield Overland Mail Route was located.

These three maps show a continuous road through northern Greene County, labeled from Boonville to Springfield, with the exception of the small area in Section 3, Township 29, Range 22. If there had been a road cut by the military in 1825, the government surveyor would surely have noted that on the map or in his field notes.

In an article published in the *Lawrence County Historical Society Bulletin*, J. Dale West discussed the route of Boonville Road through Dade County, Lawrence County, and on to Jasper County. He noted a description of "road to Boonville, Wagon Road, Boonville Road," on surveyor's notes. There is a reference to the use of "old Boonville Road" in Civil War records as it crossed Spring River east of Bowers Mill. (24)

Robert Neuman, Supervisor of Archives at the Greene County Archives, reviewed the government survey maps tracing the Boonville Road from about ½ mile west of the Spring River in Lawrence County. In Section 3, Township 28, Range 28, a road is labeled "Boonville Road." In Section 34, Township 29, Range 28, the road is labeled "road to Boonville." In Section 20, Township 30, Range 25, the road is labeled "Boonville to Fort Gibson." However, no road connection has been found on the GLO surveys from either Lawrence or Dade counties to Greene or Polk counties. It is interesting that a road is drawn across several sections and then stops with no indication of another road. Perhaps some surveyors were not as observant as others and neglected to identify road connections or the road may have been established to a particular point, after which the traveler was left to find his own route to the next connecting road.

It is possible that Nathan Boone traveled along the Boonville Road in Lawrence County and Dade County on his way to Fort Gibson. In the book, *Nathan Boone and the American Frontier*, author R. Douglas Hurt notes that Boone led his company along the trail made by other companies that had left as early as November 20, 1833. They angled southwest avoiding the village of Springfield. Apparently Boone did not want to take his troops where the inhabitants sold whiskey at 25 cents per pint. (25) The road could have been one marked on the survey as, "road to Ft. Gibson."

The focus of this book is on the Boonville Road as it traveled north from Springfield through Sections 35, 26, 23, 14 and 11 and on to the Polk County Line. This is the Boonville Road identified on early maps and the one that was used as the old Boonville mail route.

In the Greene County Circuit Court August Term 1839: "Ordered by the court that Bennet Robinson be and is hereby appointed overseer of the 2nd Division of the Boonville Road in the room of Samuel Lasley and that he call on the usual hands and that he be notified of the same. And it is further ordered that said Lasley pay all money in his hands, collected as fines, to his successor in office." (26) This is the same Samuel Lastley in whose field the surveyor set a limestone rock as a section marker and would indicate that the Boonville Road had been in use as early as 1832 when Lastley came to Greene County, and was perhaps a trail or wagon road even earlier. (See various spellings of his name in Chapter Four.)

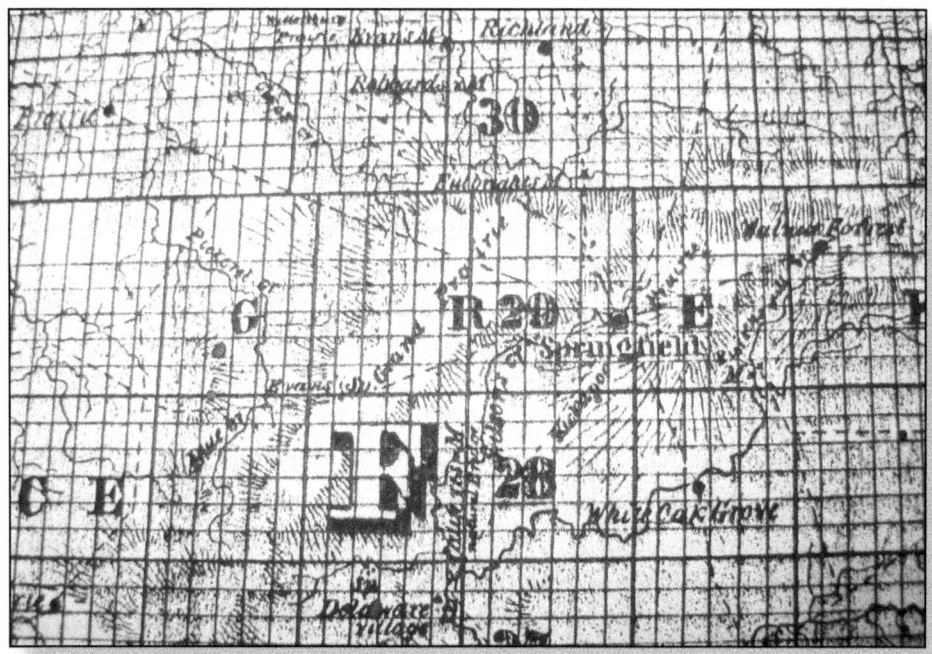

Missouri map of 1844 by Hutawa

The above 1844 map shows a road leaving Springfield, crossing two rivers at Sections 26 and 35, heading north to Richland, a post office. (27) The two rivers are Little Dry Sac on the south and Little Sac on the north.

In March 1845, an act originated in the Missouri House of Representatives to change a portion of the road leading from Bolivar to Springfield. They vacated a portion from Sections 10 and 11, Township 30, Range 22, to intersect to the old route at the north end of Joseph Evans' lane. (Joseph Evans' home was the location of Evans' Station on the Butterfield Overland Mail Route.) Plat and field notes indicated a new road to be opened must be at least 30 feet and not over 60 feet wide, and to be operated and kept up as other state roads. This act would indicate that the Boonville Road was a state road, not a county road.

On 24 February 1849, the Missouri General Assembly changed a portion of the road leading from Springfield to Bolivar "at a point where the present road makes an angle opposite the parsonage known as the end of Boonville Street." It was declared a state road and the Greene County court was instructed to keep the road opened at least thirty feet and not more than sixty feet wide and to be kept up as other state roads. The same width requirement in both these acts would indicate established standards for state roads.

Greene County Missouri County Court proceedings 4 October 1850 read: "Now at this day the Court proceeded to lay off the several roads in said County of Greene in pursuance of an Act of the Legislature of the State of Missouri approved March 10th 1849. As follows to wit: Beginning at the Court House in the City of Springfield on the Boonville Road then North to the ford of Sac River south of Tapley Daniels 1st District. Thence North to Daniel Headlees 2nd District, thence North to the Polk County line 3rd District." The ford was on the land previously owned by Samuel Lastley and sold to Tapley Daniel.

The court also laid out the Fayetteville Road. "Beginning at the Courthouse whence South to Edward Moores 1st District. Whence to the ford on Wilson's Creek 2nd District. Whence South to the Old Delaware Trace via Grand Prairie 3rd District. Whence to the county line 4th District." (28) This designation of a road from Boonville to Springfield and Springfield to the county line, and ultimately to Fayetteville, Arkansas, would complete the statute requirement for a road from the Missouri River to Fayetteville.

At the 3 October 1865 Greene County Circuit Court session, new road overseers were appointed for the county. (29) This may have been a necessary reorganization following the Civil War. A road overseer was usually a property owner along the road whose duties were to collect fines or levies and find workers to help repair and maintain the road. Alexander Evans (son of Joseph Evans, Evans' Station) was appointed overseer on Boonville Road from Little Sac River to Daniel Headlee's. Rufus Robinson (Robberson) was appointed overseer from Daniel Headlee's property to the Polk County Line. These overseers succeeded those who had been appointed by the early Greene County Court. These appointments were for the Boonville Road as previously described, so it is a puzzle why there were appointments on a Boonville/Cassville Road.

Henry Hay was appointed overseer of the Boonville/Cassville Road from Polk County to the ford of Clear Creek and John Small was appointed overseer of the Boonville/Cassville Road from Clear Creek to Lawrence County. In reviewing the 1876 Road Plat Book from Greene County, a road called Bolivar/Cassville Road began at Section 15, Township 31, Range 24 in northern Greene County, went west around Walnut Grove, then due south until it turned west at Section 5, Township 29, Range 24. (30) The appointments of Hay and Small could indicate a road connection from Polk County into Lawrence County as Mr. West had discussed. The 1879 Campbell's Atlas indicated several possible routes into Lawrence County, but nothing that went directly to Cassville, which is located south of Springfield. (31)

Although the Boonville Road was not built initially as a military road, it did acquire military significance during the Civil War. In June 1861, the second battle to secure Missouri for the Union occurred three miles east of Cole Camp, Missouri. Robert L. Owens discusses the significance of the crossroads at Cole Camp, stating that, "from the north there was Boonville Road and the Butterfield Trail which originated at the Tipton/Syracuse area." Governor Jackson waited safely in Florence until the battle was over and then passed through Cole Camp on this road going south. This allowed the governor to escape to Southwest Missouri to rally forces for the Confederacy. (32)

The map on the next page of 1862 shows a drawing of roads where U.S. Army pickets were stationed leading in and out of Springfield. The forts shown were used during the Battle of Springfield, January 1863. Boonville Road is identified in the upper left hand corner. (33)

Early Settlers Along Boonville Road

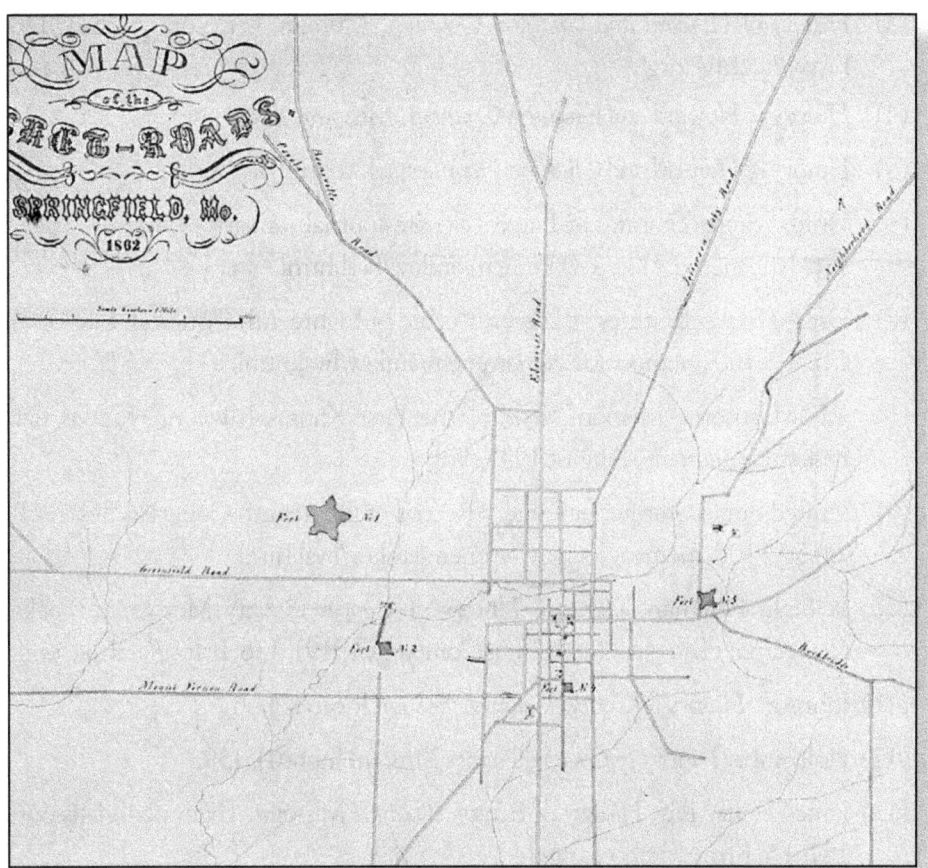

Picket's Map of 1862

The Boonville Road is mentioned several times as the armies fought in and traveled through Missouri. More skirmishes occurred around the towns of Boonville, Warsaw and Cole Camp than in Springfield, but the road was very important for troop movements throughout its length.

References
(1) Return I. Holcombe, *History of Greene County, Missouri.* (St. Louis, 1883), 25, https://archive.org/.
(2) United States Statutes at Large. "Indian Treaties, Vol. 7: 1778-1842," 107. memory.loc.gov/ammem/amlaw/lwsl.html.

(3) *History of Howard and Chariton Counties, Missouri.* (St. Louis, 1883), 126, https://archive.org/.

(4) *History of Howard and Chariton Counties, Missouri,* 127.

(5) *History of Howard and Chariton Counties, Missouri,* 167.

(6) United States Statutes at Large. "Senate Journal January 24 and 25, 1825," 106-107, memory.loc.gov/ammem/amlaw/lwsl.html.

(7) United States Statutes at Large. "Record of Eighteenth Congress, Session II, Ch.50," 100, memory.loc.gov/ammem/amlaw/lwsl.html.

(8) T.F. Morrison, "Mission Neosho, the First Kansas Mission." *Kansas City Historical Quarterly,* August 1935, Vol. 4.

(9) United States Statutes at Large. "Record of Eighteenth Congress, Session II, Ch.50," 100, memory.loc.gov/ammem/amlaw/lwsl.html.

(10) William Foreman Johnson, *History of Cooper County Missouri,* (Topeka, Cleveland: Historical Publishing Company, 1919), 138, https://archive.org/.

(11) Johnson, *History of Cooper County Missouri* [note 10], 91.

(12) Holcombe, *History of Greene County Missouri* [note 1], 159.

(13) James Henry Lay, *History of Benton County, Missouri,* (Hannibal, Missouri, 1876) 5, https://archive.org/.

(14) Lay, *History of Benton County, Missouri* [note 13], 5.

(15) Jeremiah Young, *A Political and Constitutional Study of the Cumberland Road,* (Chicago: University of Chicago Press, 1902), 89-91, https://archive.org/.

(16) Jonathan Fairbanks and Clyde Edwin Tuck, *Past and Present of Greene County, Missouri,* (Indianapolis, Indiana: A. W. Bowen Company, 1915), 40.

(17) State Historical Society of Missouri-Columbia Research Center, Columbia, Missouri.

(18) Edward M. Shepard, "Map of Indian Trails in Greene County," original map found at The Library Center, Springfield, Missouri.

(19) Fairbanks and Tuck, *Past and Present Greene County* [note 16], 43.

(20) United States Statutes at Large. "House Journal, March 1, 1836," 23, memory.loc.gov/ammem/amlaw/lwsl.html.

(21) *Ozar'Kin Vol. IX*, Number 4, Winter 1987, Ozarks Genealogical Society.

(22) Greene County Missouri Court Minutes, Book A, March 12, 1834, third day of March Term, 68–69, at Greene County, Missouri archives.

(23) U. S. Government Land Office Surveys, at Greene County, Missouri archives.

(24) J. Dale West, "The Boonville Road," *Lawrence County Missouri Historical Society Bulletin*, July 1989.

(25) Douglas R. Hurt, *Nathan Boone and the American Frontier*, permission from the University of Missouri Press, © 1998 by the Curators of the University of Missouri.

(26) Greene County Circuit Court Minutes, August 1839, at Greene County, Missouri Archives.

(27) *Missouri Map 1844, compiled from the United States Surveys and other sources*, (Edward Hutawa), at Missouri State University Library Map Department.

(28) Greene County Court Minutes October 1850, at Greene County, Missouri Archives.

(29) Greene County Circuit Court Minutes October 1865, at Greene County, Missouri Archives.

(30) Innes and Innes, *Road Plat Book of Greene County, Missouri*, (Springfield, Missouri, 1876) at Greene County, Missouri Archives.

(31) R.A. Campbell, *Campbell's New Atlas of Missouri*, (1879) at The Library Center, Springfield, Missouri.

(32) Robert L. Owens, "The Almost Forgotten Battle," http://www.colecampmo.com/civilwar/civilwar1.html.

(33) Picket's Map 1862, original map at National Archives, Washington, D.C., copy at The Library Center, Springfield, Missouri.

Chapter 2

Boonville Road and the Butterfield Mail

In 1850 California was admitted to the Union and residents began clamoring for a mail route on which mail could be delivered in less than 52 days. John Butterfield, who started out as a professional stage driver, won the $600,000 U.S. government contract to transport mail twice a week between St. Louis and San Francisco in 25 days or less.

The railroads, fearing they could not compete with steamboats, had earlier decided to locate the rails away from the Missouri River. By 1858 the westernmost extension of the railroad had been built to Tipton, Missouri, located just ten miles east of the old Boonville mail road. This fit into Butterfield's plan because the mail could be brought by train from St. Louis and, in Tipton, loaded on stages bound for California. This road became the route of the Butterfield stage through Missouri. Butterfield's work was relatively easy in Missouri because of the already established roadbed. When the route went further west, he had to clear roads, build bridges and ferries and deal with hostile Indians.

In addition to the Overland Mail Route, Butterfield's company established mail and passenger coaches between Springfield and Tipton. They were reportedly always crowded to capacity. (1)

The first successful run was completed from Tipton, Missouri to San Francisco, California, September 16–October 10, 1858. J. Waterman Ormsby, a reporter for the *New York Herald*, was the first traveler on this 2800-mile route and his oft-quoted narrative of the trip provides insight into location of stations and conditions of travel.

A man named Albert D. Richardson also wrote about his experiences on the

Butterfield Mail. In his book, *Beyond the Mississippi*, he wrote:

"On the fifteenth of August, [1859] I ......started for the far frontier. At Syracuse, one hundred and sixty-eight miles west of St. Louis, and then terminus of the Missouri Pacific Railway, I left the cars for a coach of the Butterfield Mail Company."

"Our coach, leaving Syracuse after dark, jolted along for fifty miles during the night, and at sunrise stopped for breakfast in Warsaw, Benton County—a genuine southern town......We forded the Osage though it is navigable above Warsaw for half the year. The region was hilly and rocky, intersected by many streams and timbered with a dozen varieties of oak; the houses long and low with outside chimneys... After passing some beautiful prairies and enduring another night of uneasy slumber, we woke in Springfield, on the summit of the Ozark Mountains— the leading town of southwestern Missouri. ... Springfield had pleasant, vine-trellised dwellings and two thousand five hundred people. The low straggling hotel with high belfry, was on the rural southern model: dining room full of flies, with a long, paper-covered frame swinging to and fro over the table to keep them from the food; the bill of fare, bacon, corn bread and coffee; the rooms ill-furnished, towels missing, pitchers empty, and the bed and table linen seeming to have been dragged through the nearest pond and dried upon gridirons." (2) He noted that Missouri had unequaled resources and grand rivers and was prosperous and flourishing. However when he passed through two years later, he found "the country blazing with civil war which swept away many fruits of the labor of twenty years..." (3)

Roscoe and Margaret Conkling's book is considered a prime source for information about the Butterfield Mail. Their book traces the route from Tipton to California.

Following is the map from Conkling and Conkling showing the Butterfield Overland Mail Route through Missouri. (4)

The stations which Butterfield located between Tipton and Springfield were: Shackleford's, Mulhollen's, Burns', Cole Camp, Warsaw, Bailey's, Bolivar, Yoast's, Smith's and Evans'. Some of the locations have been difficult to identify because of changes in roads and the construction of dams on the Osage and Pomme de Terre rivers. Don Matt, in his excellent blog has pictures of the current locations of these stations, which he found while traveling on his motorcycle. (5) Some of the towns

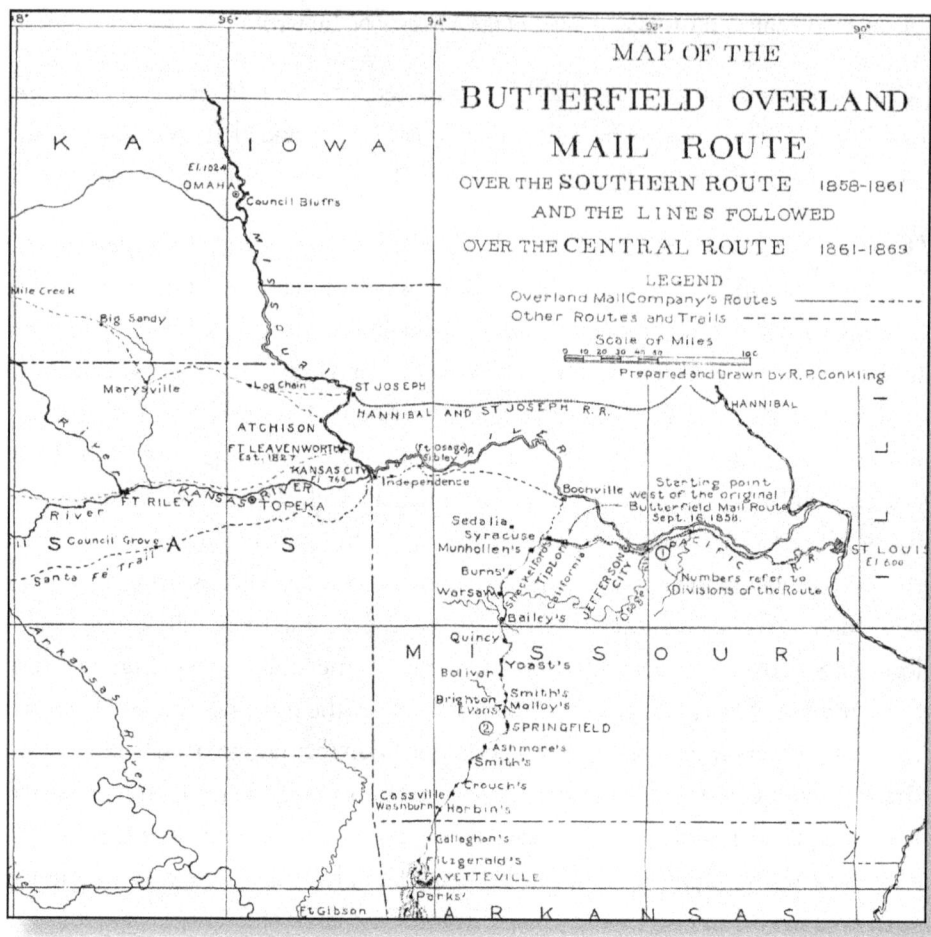

Butterfield Overland Mail Route through Missouri

have erected signs to identify that the Butterfield Overland Stage passed through their town.

My focus is on Section 14, Township 30, Range 22 in Greene County, where Evans' station was located, and on sections located south of Section 14, but still on the Boonville Road.

The last station before the stage entered Springfield was Evans' station. That's just two sections north of the David Murray farm, which is located in Sections 26 and 35. As Conkling described, "From Evans' the mail road continued on almost due south to the crossing on Little Sac River about three and a half miles south of the station. Three quarters of a mile farther on it crossed South Dry Sac Creek, and passed through the forgotten little settlement of Fair Play, and then curved slightly southeast for a distance of about four miles and entered the city of Springfield, the station nine miles from Evans', over the old Boonville Road, now Boonville street, which it followed to the public square."(6)

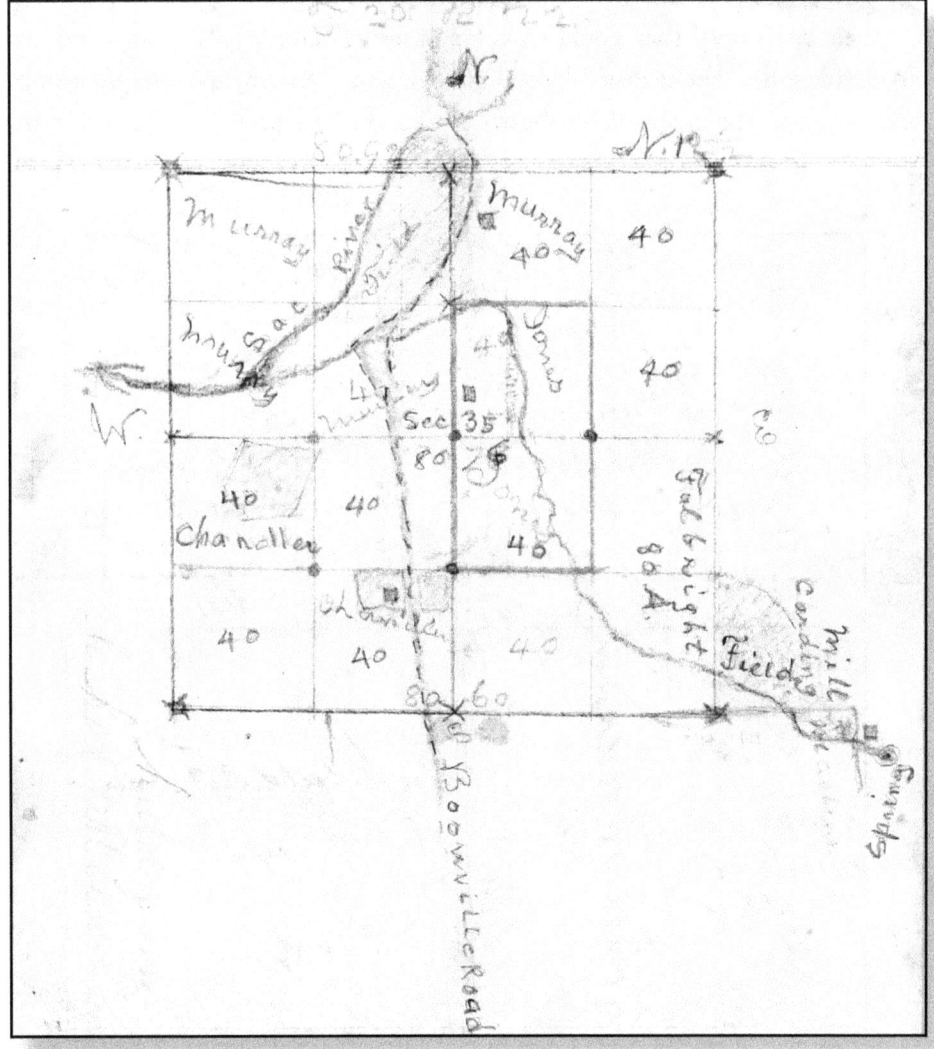

David Murray Map 1867

This description of the route between the two Sac Rivers fits exactly the route outlined in the David Murray map, which must have been given to him when he purchased the farm in 1867. (7)

Note the field, which is in the same location as Mr. Lastley's field on the 1835 government survey report, the two rivers, and what looks like two fords on the south Little Dry Sac.

The U.S. land office did not open in Springfield until 1835 and Holcombe states that there were only 90 cash entries made the entire first year. (8) Settlers squatted on their lands until they could make application, have surveys completed and finally have the patent issued by the government. This process sometimes took several years. The patent dates shown in Section 14 reflect the dates when the

Settlers in Section 14, Township 30, Range 22

applications were made and in some cases the settlers sold the land before they received the actual patent. The earliest patentees in Section 14 were Henry Morrison, Archibald Adams, Thomas Wilson, Theophilus Leathers and Samuel Lastley. Most of the early patentees in Sections 14, 23, 25, 26, 35 were from counties in Tennessee, coming to Greene County in the early 1830s. The family relationships of those living in Section 14, where Evans' station was located, are detailed below. Plat Information is found in *An Index of the Springfield Land Office Sales Book, 1833-1892*. (9)

Before exploring family relationships of the early patentees in Section 14, it is important to learn about some of the other settlers on adjacent sections, because they were related to each other. According to Holcombe (10) John Headlee came to Greene County with his brothers-in-law, Benjamin Johnson and James Dryden in 1832. Johnson and Dryden settled on the Little Sac River. They were brothers-in-law to Bennett Robberson, Zachariah Sims, Henry C. Morrison and Robert Sims.

Elizabeth Pettigrew Robberson brought her large family from Bedford County, Tennessee. She and her sons entered land patents in Sections 3, 10, 11, Township 30, Range 22, for a total of almost 1000 acres. This land became known as Robberson Prairie and in 1837 Robberson Prairie Township was created with the election precinct located at the house of Elizabeth Robberson. Her son Bennett was named earlier as a road overseer by the Greene County Court, succeeding Samuel Lastley, mentioned in Chapter One.

The brothers-in-law named above were married to daughters of Fanny Sims, who moved with her family from Bedford County, Tennessee, probably in 1833. In Marsha Rising's book, the family relationships are detailed in a Bedford County deed following the death of Fanny's husband, Briggs. Twelve children listed were: Washington Sims, Briggs Sims, Zachariah Sims, John Sims, Burwell Sims, Holly Sims, Delphia Sims, James Dryden and his wife Frances, formerly Frances Sims, John Hadley (Headley) and his wife Polly, formerly Polly Sims, Benjamin Johnson and wife Sally Johnson, formerly Sally Sims, Bennet Robertson (Robberson) and wife Elva D. Robertson, formerly Elva D. Sims and Clinton Morrison and wife Nancy, formerly Nancy Sims. (11)

With that background, it may make more sense to learn about the first patentees in Section 14.

Henry Clinton Morrison was born sometime between 1800-1810, possibly in Bedford County, Tennessee because he married Nancy C. Sims, daughter of Briggs and Fanny Sims in Bedford County. Henry was in the 1830 census of Bedford County and left the county after 1832 when his father-in-law's heirs were named, one of whom was Nancy C. Morrison. The patents on his land in Section 14 were issued 10 April 1843 and 10 September 1844, but were applied for in 1838 and 1840. He and his family were listed in the 1840 Greene County census.

Henry Morrison sold his 120 acres, obtained by land grants #1243 and #5034 to Thomas Wilson, 3 February 1841. (12)

Thomas C. Wilson was born 22 November 1800, in Hardeman County, Tennessee. His wife was Margaret Bond, born 8 November 1800. Thomas C. Wilson first appeared in the 1834 Greene County tax list. He and his family were listed in the 1840 Greene County census. Neighbors listed that had family connections that follow in the narrative were William Robertson (Robberson) and Hosea Mullings.

In May 1842, Thomas Wilson wrote a will leaving his plantation to his wife. He said the Morrison place on Boonville Road was to be sold first to benefit the estate. (13) He named friends Hosea Mullings and Elisha Headlee as executors. (Note: In the early records, Headlee may be spelled Headlea, Headley). In May 1842, Mullings and Headlee bought the property from the estate in a private sale. (14) Then in 1847, Mullings and Headlee sold the SW ¼ of the NW ¼ Section 14, Township 30, Range 22 to Samuel Lee. (15) There was mention of a will by Samuel Lee in Greene County, but no will was found in the Greene County Archives.

Archibald Clinton Adams was born 4 July 1780 in North Carolina. He married Frances McClure Dryden (sister to James Dryden) about 1810 in Bedford County, Tennessee. He was listed in both the 1820 and 1830 census in Bedford County, but in 1840 he was listed on the census in Greene County, Missouri. He acquired by patent 480 acres in Greene County, 320 acres of which were located in Section 14. Archibald wrote his will 4 June 1844, naming his eldest daughter Mary Ann and her husband Robert Sims, who were to have 113 acres in Bedford County,

Tennessee; daughter Eliza A. and her husband Zachariah Sims were given 160 acres in Greene County; Nancy and her husband Samuel G. Headlee received 160 acres in Greene County (16). Robert and Zachariah Sims are sons of Fanny and Briggs Sims of Bedford County, Tennessee.

Samuel Lastley is shown as the owner of the SW ¼ of the SW ¼ of Section 14. His history is covered in Chapter Four, Early Settlers in Section 26 and 35. It is not believed that he lived in Section 14. The fact that he bought land in Section 14 raises the possibility that his wife, Charity Johnson, might have been a sister to Benjamin Johnson. No proof has been found, but the families that came from Bedford County and Maury County, Tennessee all arrived in 1832 and 1833, so this relationship is possible.

Theophilus Leathers applied for a patent on the NW ¼ of the SW ¼ of Section 14, Township 30, Range 22 in November 1840. The patent #5472 was issued 5 May 1845. At the same time he applied for a patent on the SE ¼ of the NW ¼ of Section 14, Township 30, Range 22. Patent #5473 was issued 5 May 1845. This was the same property that Leathers sold to Joseph Evans in March 1848. (17) This property became the location for Evans' Station.

A relationship exists between Joseph Evans and Theophilus Leathers, although the exact relationship is not known. In North Carolina Marriage Bonds 1741-1868, Theopheleys Leathers married Elizabeth Panels 14 May 1836 in Davidson County, North Carolina. Bondsman was Joseph Evans. Evans and Leathers both state in census reports that they were born in North Carolina.

In the 1840 census of Rutherford County, Tennessee, Joseph Evans is listed as head of the family with 1 male under 5, 1 male 5-10 years, 1 male 30-40, 2 females 5-10 years, and 1 female 30-40. These ages would correspond with his children, John, Alexander, Melinda and Eliza. The 1840 census of Rutherford County also lists Theophilus Leathers as head of family with 1 male 20-30, 1 female under 5, and 1 female 15-19. This would suggest that the two families were enumerated in June 1840, moved to Greene County shortly after and began purchasing land in November 1840. Joseph Evans purchased 40 acres in Section 24 and 40 acres in Section 25 and received the patents 5 May 1845.

Multiple entries on Ancestry.com state that Joseph Evans' wife was Elizabeth Leathers, but no documentation has been provided and a marriage record has not been found. She could have been a sister to Theophilus. Records showing that Joseph provided a marriage bond for Theophilus Leathers; that the families were found living in Rutherford County, Tennessee; and that they had real estate transactions together in Greene County could indicate that Theophilus and Elizabeth Leathers Evans were brother and sister.

On the 1850 census, Robberson Township, Greene County, Missouri, Theophilus stated that he was 35 years of age, a farmer, born in North Carolina. His children were John age 13, Eliza age 11, Edward age 9, Nancy age 7 and Sarah age 2. He died 21 March 1857 and was buried in the Robberson Prairie Cemetery, located on the Robberson Prairie, named after Elizabeth Robberson who settled there with her sons in 1832.

In the 1880 census of Robberson Township, Theophilus' widow Elizabeth was living with William and Nancy Robberson, where she is listed as mother-in-law. William is the son of Bennett and Elva Robberson. Note: in records the Robberson family name is variously spelled Robberson, Roberson, Robinson and Robertson. Robberson Township was spelled Robinson by some census takers. Robberson appears to be the correct spelling.

Joseph and Elizabeth Evans' children married the children of neighbors. John, their eldest, married Mary Wilson, daughter of Thomas Wilson. Alexander Evans married Jane Caroline Robberson, daughter of Elizabeth Robberson. Daniel McCord Evans married Mary Evelyn Robberson, daughter of Allen Robberson. Eliza Elizabeth Evans married George Cathey Mullins, son of Hosea Mullins. Joseph and Elizabeth Evans are both buried at Robberson Prairie Cemetery.

Joseph Evans has been described as a prosperous and influential citizen of early Greene County. He built both a sawmill and a gristmill and constructed his home from walnut logs found on his property. Conkling and Conkling described the house: "The original house and station was a two story, dressed walnut log building, with porches and red brick chimneys. From a survey of the ruins, the approximate dimensions of the building were thirty seven feet six inches long by thirty six feet six inches wide." (18) This home would have been a suitable selection for a Butterfield way station. In it, Evans could have fulfilled these stage stop requirements: Take care of property and animals and be accountable for safety of passengers and mail and keep horses and mules ready for the road at all times. It was clearly a job for a responsible person.

The location of the old Evans' station is east of the north-bound lane of Highway 13, and a little north of the dead end of Farm Road 141. The location is on private property and cannot be accessed by the public. Permission was obtained to visit the site and take photos in 2008. The picture, below, shows remnants of the old fireplace; there are also cornerstones showing where the house was located. Looking south, the tracks of the old Boonville Road are faintly visible.

Don Matt pinpoints the Evans' station location in his blog. His story can be found after accessing the blog site and scrolling to the entry, "Evans' Station Update 2008." His narrative provides interesting and informative reading about the Butterfield Mail Route through Missouri. (19)

Remnants of a fireplace at Evans' Station

Photo of the Evans' Station property looking north. Photo by Don Matt 2008

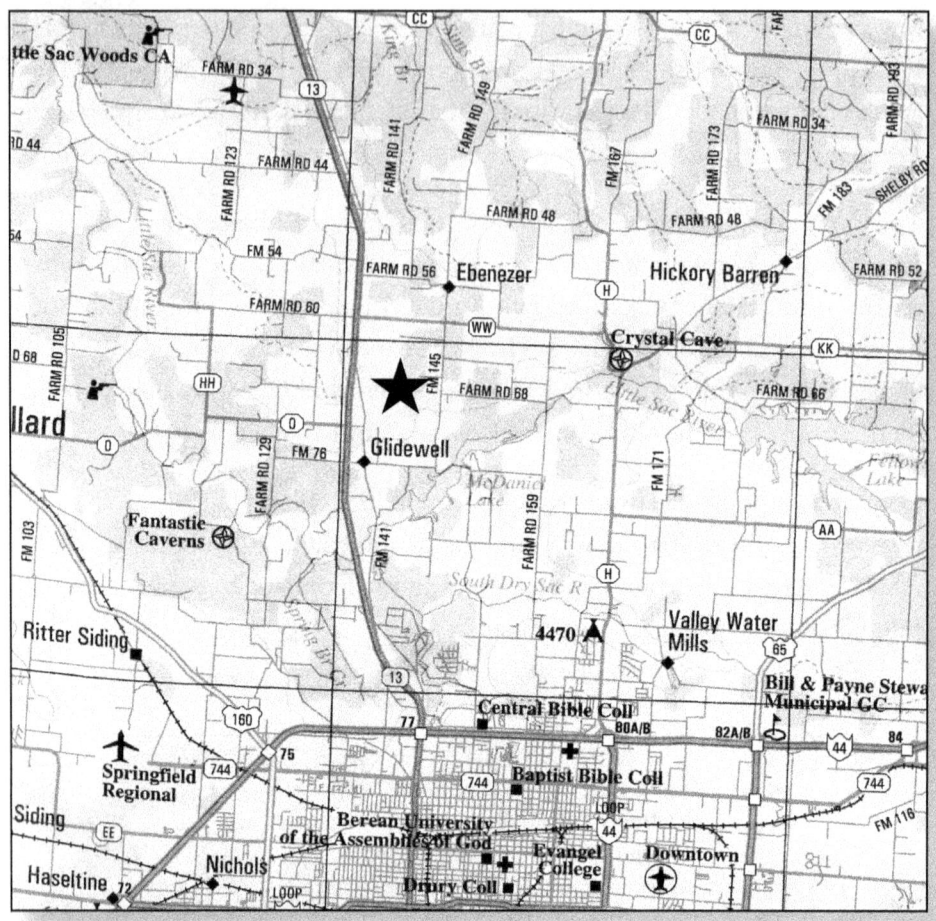

The 1998 Missouri Atlas and Gazetteer, *reprinted with permission.*
© 2010 DeLorme (www.delorme.com) *Delorme Missouri Atlas and Gazetteer*

Just as there was confusion over the naming of the "Boonville Road, Old Osage Trail, road cut by the government in 1825," there has been confusion among researchers about where Evans' Station was located. This is probably because the roads have been re-located and renamed.

The original Boonville Road, which passed in front of the David C. Murray home, was moved further west in 1928 and named Highway 13. (See Chapter Three). The old Highway 13 (also called Bolivar Road) is now named Farm Road 141. In 1975, Highway 13 was moved still further west and improved to a four-lane highway. In looking at the maps, it is difficult to place the Evans' Station without knowing how the old Boonville Road became Bolivar Road, then Highway 13, and finally Farm Road 141.

The 1998 Missouri Atlas and Gazetteer shows Farm Road 141 ending at the new Highway 13 just north of Glidewell, and north of Highway O. The Evans' Station is to the east, very close to the star indicating woods. There is no road access to the property. (20)

From the Evans' station, the Boonville Road traveled south through Section 23 where early settlers Ezekiel Cook, Alexander and John Evans (both sons of Joseph) and Hosea Mullings lived.

Hosea Mullings (Mullins) had a lengthy history in Greene County. He, too, was from Bedford County, Tennessee. One source indicates that he came to Greene County in 1818; however, because Hosea was listed in the 1830 census of Bedford County, it is more probable that he came to Greene County around 1832-33. He first appears in the 1835 tax list in Greene County. His son, George Cathey Mullins married Eliza Elizabeth Evans, daughter of Joseph and Elizabeth Evans, on 27 March 1856, Greene County, Missouri. *The Pictorial and Genealogical Record of Greene County* stated that in the early days, Hosea "did considerable teaming from Boonville for Springfield merchants, and while thus employed experienced a good many hardships."(21)

Boonville Road continued through Section 26 where early patentees Tapley Daniel and Samuel Austin owned property, and through Section 35, properties owned by early patentees Arthur Davis, Samuel Lastley, Tapley Daniel, Silas Baker and William Fulbright. Information on early patentees and settlers in these sections, as they relate by intermarriage and business, will be covered in Chapter 4, Early Settlers of Sections 26 and 35.

The Colton sectional map of Missouri, published 1869, shows the Boonville Road continuing south through Section 3, Township 29, Range 22. (22) The map indicates it passed by a settlement and mill at the town of Fair Play, as noted by Conkling and Conkling. Research has not revealed the significance of Fair Play. There is some confusion with the name because another town named Fair Play existed in Polk County at the same time.

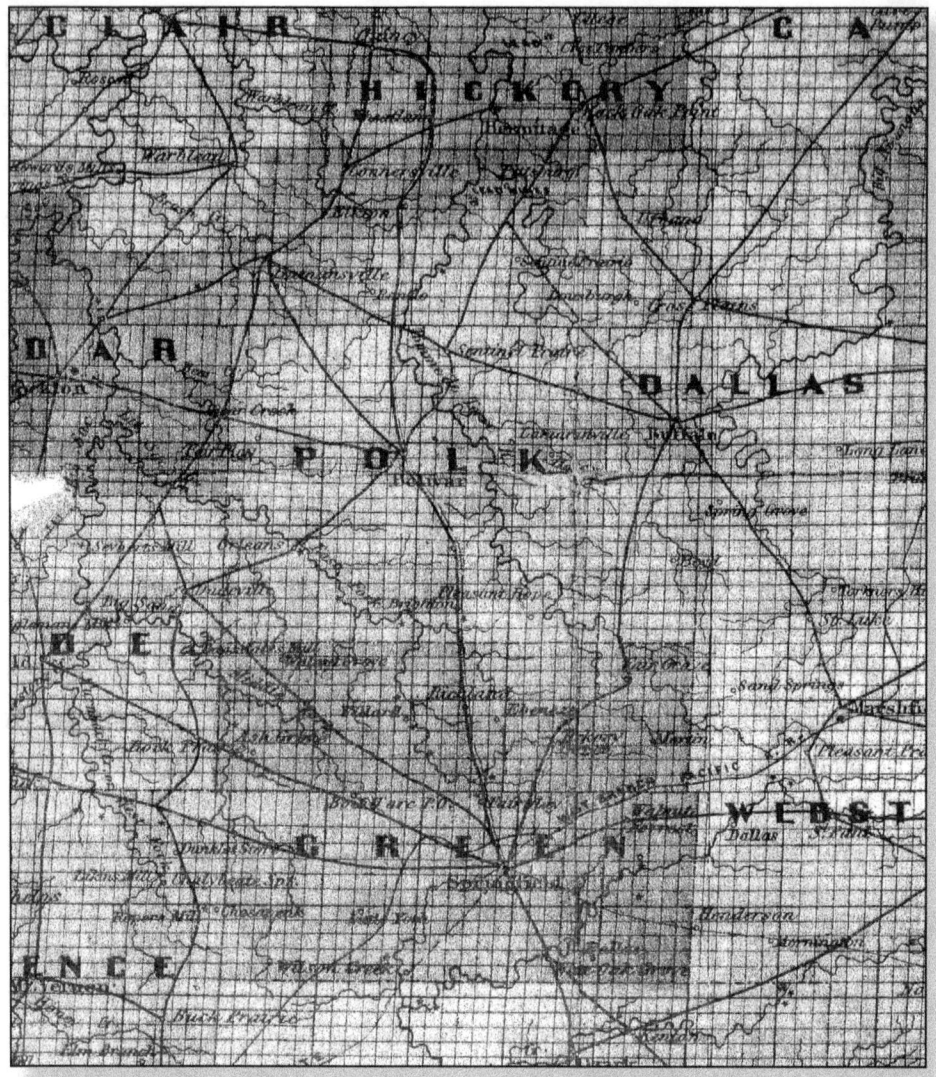

Colton's Map 1869

Lloyd's Military Map of 1861 (23) also shows Fair Play and indicates a proposed branch of the Southern Pacific Railway passing north of Springfield and very close to Fair Play. During the decade 1860-1870, there may have been plans to locate the railroad closer to Fair Play and by-pass Springfield; however, the city leaders were determined to locate the railroad in the city and in April 1870, the Pacific Railway train chugged into North Springfield.

Lloyd's Military Map 1861

Colton's Railroad and County Map of the Southern States 1864 shows approximately the same route as Lloyd's Military Map. After the railroad was completed to Springfield, the little town of Fair Play is no longer shown on maps. Colton's 1876 map shows the railroad going through Strafford and Northview before arriving in Springfield and Fair Play has disappeared.

After traveling through the town of Fair Play, the Butterfield Overland Mail route into Springfield is subject to conjecture. If the 1862 map of picket locations is correct (see page 17), it would indicate that Boonville Road ran into Springfield along what is now known as Fort Street to the Greenfield Road (now High Street), turned east and went to Boonville Avenue, turned south and went directly to the public square stopping at Smith's Tavern. (This is the stop described earlier by Mr. Richardson.) There are only two bridge crossings noted on the picket map, one on Boonville Avenue and one on Jefferson Avenue. The crossing on Jefferson Avenue would have been too far east to be practical.

The 1883 *History of Greene County* states that in 1837 the county court appropriated $100 to build a bridge "across the town Branch, north of the public square, at Springfield." (24) This would be the bridge across the Jordan Creek on Boonville Avenue. It would make sense that Butterfield would use this route to the Smith's Tavern stop because his goal was to deliver the mail in the fastest time possible. A recollection from Hubble Personal Reminiscences at the Old Settlers Reunion said, "---the first Overland coach arrived in Springfield—the horses came up Boonville hill at a gallop." (25)

In 1860 there were few platted additions to the city of Springfield. Those that were platted were in the area of downtown plus extensions down Main Street close to the present Drury College (Lynn, Calhoun, Webster, Scott and Nichols streets) and along Jordan Valley. If the Butterfield Stage took another route into the city, it would have required jogging from say, Campbell or Grant avenues over to Boonville.

There is an interesting description about the lack of development north of Springfield written by a reporter for the *New York Times*, 26 January 1863. He described the Battle of Springfield, the location of forts and defense of the city. He said, "The city thus fortified lies half in the prairie and half in the timber. Upon the north and east all is forest; upon the south and west the country is entirely open." This could indicate that there were roads through the timber that do not correspond to current streets.

In the summer of 1958, the Missouri Historical Society announced that the organization would mark stage stations with signs commemorating the 100th anniversary of the Butterfield Overland Mail. The Greene County Historical Society assisted in identifying the route in Greene County by placing markers at Evans' Station on the old Highway 13 (close to the original site), near the present U.S. Medical Center on Sunshine Street and in front of the Lee Phillips farm home on Highway ZZ. The Greene County Historical Society also erected another marker near the Dickerson Park Zoo, close to the point where the stage entered the present city limits. This marker was moved when the four-lane Highway 13 and the 1-44 Interchange were constructed. It was re-located adjacent to the food pavilions at Dickerson Park Zoo, opposite Fort Street. Since the construction of 1-44, Fort Street dead-ends near this location.

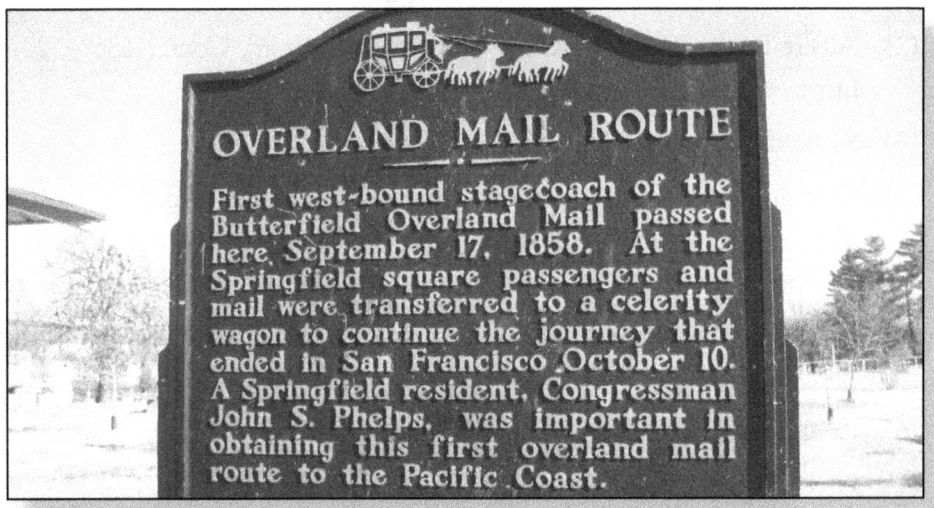

Overland Mail Route sign at Dickerson Park Zoo

Records of the Greene County Historical Society available at the Missouri State University Library indicated that the committee formed to locate the signage relied heavily on maps from the US Geological Survey Maps of 1951. The results were compiled in Kingman's *Maps of the Butterfield Overland Mail through Missouri and Arkansas*. (26) Unfortunately the route shown leading into Springfield is only a curved line superimposed over current streets.

The Historical Society marker indicating the location of the Evans' Station is no longer standing. Farm Road 141, which was the route of the old Highway 13, dead ends very close to where the marker was located. Contact was made with the Missouri Department of Transportation to find what had happened to the marker when the new four-lane Highway 13 was constructed in 1975. There were no records available to indicate if the marker was removed by a collector or destroyed when the highway was built.

References

(1) Roscoe Conkling and Margaret Conkling, *The Butterfield Overland Mail 1857-1869*, (Glendale, California: The Arthur Clark Co., 1947), 182.

(2) Albert Richardson, *Beyond the Mississippi*, (Hartford, Conn., 1869), 207, https://archive.org.

(3) Richardson, *Beyond the Mississippi*, 209.

(4) Conkling and Conkling, *Butterfield Overland Mail* [note 1], Vol.3, Map Section, Sheet #1. Reprinted with permission; University of Oklahoma, owner.

(5) Don Matt, "Butterfield Overland Mail, 2000 Miles of Motorcycling the Butterfield Trail," *butterfieldoverlandmail.blogspot.com*. 2006.

(6) Conkling and Conkling, *Butterfield Overland Mail* [note 1], 180.

(7) Private papers, David C. Murray.

(8) Return I. Holcombe. *History of Greene County Missouri*, (St. Louis, 1883), 177, https://archive.org.

(9) *An Index of the Springfield Land Office Sales Book 1833-1892*, Greene County Archives Bulletin #36, (Springfield, Missouri: Greene County Archives).

(10) Holcombe. *History of Greene County Missouri* [note 8], 148.

(11) Marsha Hoffman Rising, *Opening the Ozarks: First Families in Southwest Missouri 1835-1839*, (Derry, New Hampshire: American Society of Genealogists, 2005), 1986. ISBN#1-59975-350-2.

(12) Greene County Missouri Deed Book B-126.

(13) Greene County Missouri Will Book C, No. 2:11-13.

(14) Greene County Missouri Deed Book C-210.

(15) Greene County Missouri Deed Book E-1.

(16) Greene County Missouri Will Book C:35-38.

(17) Greene County Missouri Deed Book E-412.

(18) Conkling and Conkling, *Butterfield Overland Mail* [note 1], 179–180.

(19) Matt, *butterfieldoverlandmail.blogspot.com*. Update 2008.

(20) *Missouri Atlas & Gazetteer*, (Yarmouth, Maine: DeLorme), 1998, 52.

(21) *Pictorial and Genealogical Record of Greene County, Missouri.* (Chicago, 1893), 69.

(22) *Colton's New Sectional Map of Missouri*, (New York, 1869), copy at Missouri State University Library Map Department, Springfield, Missouri.

(23) *Lloyd's Great Military Map of the Fifteen Southern States*, (New York, 1861), copy at Missouri State University Library Map Department, Springfield, Missouri.

(24) Holcombe, *History of Greene County Missouri* [note 8], 183.

(25) Martin Hubble, "Personal Reminiscences and Fragments of Early History of Springfield and Greene County Missouri," Springfield Sesquicentennial Edition, 1979, at Greene County Missouri Archives.

(26) W. A. Kingman, *Maps and Pictures of the Route of The Butterfield Overland Mail Thru Missouri and Arkansas*, (1958) at The Library Center, Springfield, Missouri.

Chapter 3

Boonville Road becomes Bolivar Road

Because John Butterfield's debts to Wells Fargo were so extensive, he was forced out of the stage company in 1860 and the company assets were transferred to Wells Fargo. On 21 March 1861, the U.S. Congress revoked the mail contract in anticipation of the coming Civil War. Those factors, plus competition from the railroads, put an end to the southern route of the Butterfield Overland Mail.

Each time the railroad stopped building, a boomtown sprang up only to bust when the rail lines were completed to the next section. In the *History of Pettus County, Missouri*, author Demuth I. MacDonald remembered: "In 1860, the then thriving railroad termini, first Tipton, then Syracuse, then Otterville and Smithton, were crowded with wagons of every description, and with goods in immense quantity, brought there by railroad, and taken thence by wagon trains to the west and southwest. As soon as the railroad left each of these towns, the big houses and the great trade, the hum of excitement of business left them, and rolled on, like a tide, with the thunder of the iron horse, to the next terminus." (1)

Levens, in his *History of Cooper County* described the coming of the railroad to Otterville. "Otterville commanded quite a brisk trade, presented a very active and businesslike appearance, and indeed for a time it flourished like a 'green bay tree.' But it was not destined to enjoy this prosperity long. The railroad company soon pulled up stakes and transferred the terminus to the then insignificant village of Sedalia….it soon rose like magic, from the bosom of the beautiful prairie, and in a few years Sedalia has become the county seat of one of the richest counties in the state, and a great railroad centre, while truth compels me to say that Otterville has sunk back into its original obscurity." (2)

The railroad reached Sedalia in 1861 and Sedalia became the supply depot for overland provisions just as the previous towns had been. Wagon trains came from Southwest Missouri as well as Kansas, Indian Territory, Arkansas and Texas to purchase their goods. However, further construction on the railroad was delayed until the end of the Civil War.

Sometime around the end of August 1867, David C. Murray and his wife Huldah, his daughter Ezenith and their sons, Jasper, Zelotus, Andrew and William boarded a train in Upper Sandusky, Wyandot County, Ohio, bound for Greene County, Missouri. Jerema Sell Dixon, daughter of Ezenith, remembered her mother telling that in the hubbub of transferring animals and property to another train in St. Louis, the family lost little Andrew and they feared they would have to go on without him. Luckily he was found in time to continue on with the family to Sedalia. After leaving Sedalia, they traveled for five days with their wagon and animals, hiding at night to avoid bushwhackers, who were still very active in the Ozarks following the Civil War.

By the spring of 1870, the southern line of the Southwest Pacific Railroad was completed to Springfield. Now commerce was centered on St. Louis, not Boonville or Sedalia, and it was shortly after this that the Boonville Road became known as the Bolivar Road. In the 1876 Greene County Missouri Plat Book, the Boonville Road was identified as the Bolivar Road. (3)

In April 1889, the citizens petitioned the Greene County Court "to establish two suitable bridges across Sacs River at a point where the Ebenezer and Pleasant Hope road crosses the said stream, which point is a few rods east of the crossing of the Springfield to Bolivar road." They state that: "We would call your attention to the fact that the citizens of the northern portion of the county suffer great inconvenience on account of the frequent impassable conditions of said stream in coming to the County Seat." (4) The petition was signed by D.M. Evans, T.S. Wilson, B.H. Robinson, Sam Headlee and James Ross, who were sons of early settlers.

It is unclear whether they want one or two bridges erected at this location, but this is made clear in a later road change to Ebenezer, which will be shown in a 1904 document. It makes one wonder if the Butterfield Overland Mail experienced the same difficulties in crossing Sac River located in Section 26.

Apparently the Greene County Court agreed that a bridge was needed, for the court advertised for bids to build the bridge. The Greene County Court, 4 November 1889, rejected a bid by the Wrought Iron Bridge Company of Canton, Ohio, to erect an iron bridge over Sac River. (5) The bid of $3650 was considered too high and the project was re-advertised.

On 27 November 1889, the Wrought Iron Bridge Company revised their bid to $2995 and they were awarded the contract. On 20 June 1890, the Court accepted the bond of $6000 from the Wrought Iron Bridge Company for the building of sub- and super-structure of an iron truss bridge. (6) The bridge was completed in June 1890.

Farmers need to keep their cattle from wandering away and rivers are usually fenced in by "water gaps," which is any kind of crude fencing that will keep the animals from heading down river. This usually works until there is a flood, when the gaps break, float away and must be built again. In July, 1890, Z.G. (Zelotus) Murray, son of David Murray, signed a contract with the county court allowing him to attach or suspend water gates from the lower chords of said bridge and described the location of the bridge as being on the old Bolivar Road in Section 26, Township 30, Range 22. (7)

In August 1896, citizens again petition the court for a bridge, this time over Little Dry Sac. The petition reads: "We the undersigned free holders of Robberson, Campbell and Franklin Townships, said County and State, we would respectfully petition your Honorable Body to erect a bridge over the Sac Creek on the Bolivar Road just East of and adjoining the present ford and on the land of Z. Murray in Section 35, Township 30, Range 22." (8)

In April 1902, the residents of Robberson Township again petition the Greene County Court for a bridge over Sac River. They ask for "A bridge over Sac Creek where the same crosses over Bolivar Road, being near the southeast corner of the southwest quarter of Section 26, Township 30, Range 22." (9) This request appears to be in conflict with the 1889 request, which resulted in a bridge built over Sac River in 1890. Citizens appear to be requesting a second bridge on Bolivar Road. There was a ford that residents crossed on their way to Ebenezer and is shown in the 1876 Greene County Plat Book. Apparently residents wanted a bridge built at the location of the ford.

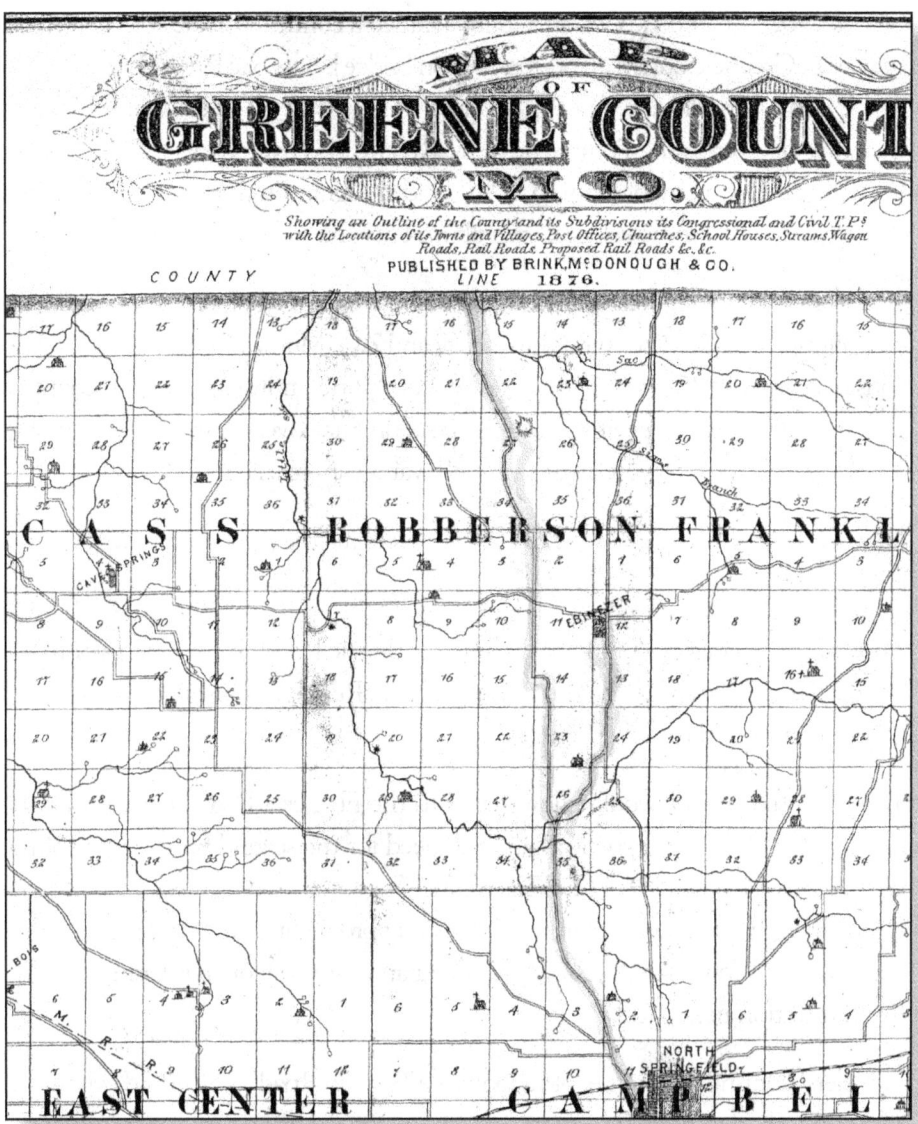

1876 Greene County Plat Book

The map above shows the Boonville Road on the left and the road to Ebenezer on the right. The road to Ebenezer turns just before crossing the Little Sac River. (10) There is still a ford on the Murray farm, which must have been the ford identified on the above map.

10 July 1902, the Greene County Court awarded a contract for the construction of bridges in Greene County: an 80-foot wagon bridge across Sac River on the Bolivar Road, an 80-foot bridge across Sac River on the Pleasant Hope Road and a 75-foot wagon bridge across the Pomme de Terre River east of Fair Grove. The Wrought Iron Bridge Company of Canton, Ohio, bid the grand sum of $4078.50 for the construction of all three bridges, their abutments and superstructures.

The Bolivar Road reference above is to the bridge on Little Dry Sac River, which is confirmed by a deed between the county and Z.G. Murray. On 30 April 1903, in consideration of $25 paid by the county, Z.G. Murray transferred to the county a strip of land at the bridge at the Murray Ford on Sac River, Section 35, Township 30, Range 22, reserving the right to attach water gaps to said bridge. (11)

Now we solve the puzzle of the road to Ebenezer. See the diagram on the following page.

This shows the location of the bridge over Little Sac River and an exit prior to the bridge, the same as shown on the 1876 Plat Map. Residents must ford the river and proceed on the old road to Ebenezer.

The proposed change would allow access to the Ebenezer Road *after* crossing the bridge, thus eliminating the ford and the need for the second bridge. Date of the change was 14 July 1904. (12) This road change was used until Greene County closed the road permanently in the 1990s at approximately the northeast quarter of Section 26, Township 30, Range 22. The part that is still open is Farm Road 143 [see illustration next page].

Travelers continued on the old Boonville/Bolivar Road until 1928 when the Missouri Department of Transportation eliminated the curved section between the rivers and the road no longer passed in front of the David Murray house. The name Bolivar Road was discontinued and the road was re-named Highway 13. After the four-lane Highway 13 was built in 1975, the name changed from Old Highway 13 to Farm Road 141. The embankment approach and one girder are still standing where the 1890 bridge was located in Section 26 and part of the embankment is visible in Section 35.

(The Commissioner will state whether, in his opinion, the proposed Road will be of great public utility or not.)

The proposed change of Road will be of Pu utility, by giving access to the Bridge on S

All of which is respectfully submitted. E. J. Rhodes .... Co. Su
And Ex-Officio Commissioner of Roads and Br

Tp. 30   Range 22

26 change
25
Bolivar Road
Old Road to Elkmon
Sac River
Bridge Road

July 14 1904

Diagram of proposed road change 1904

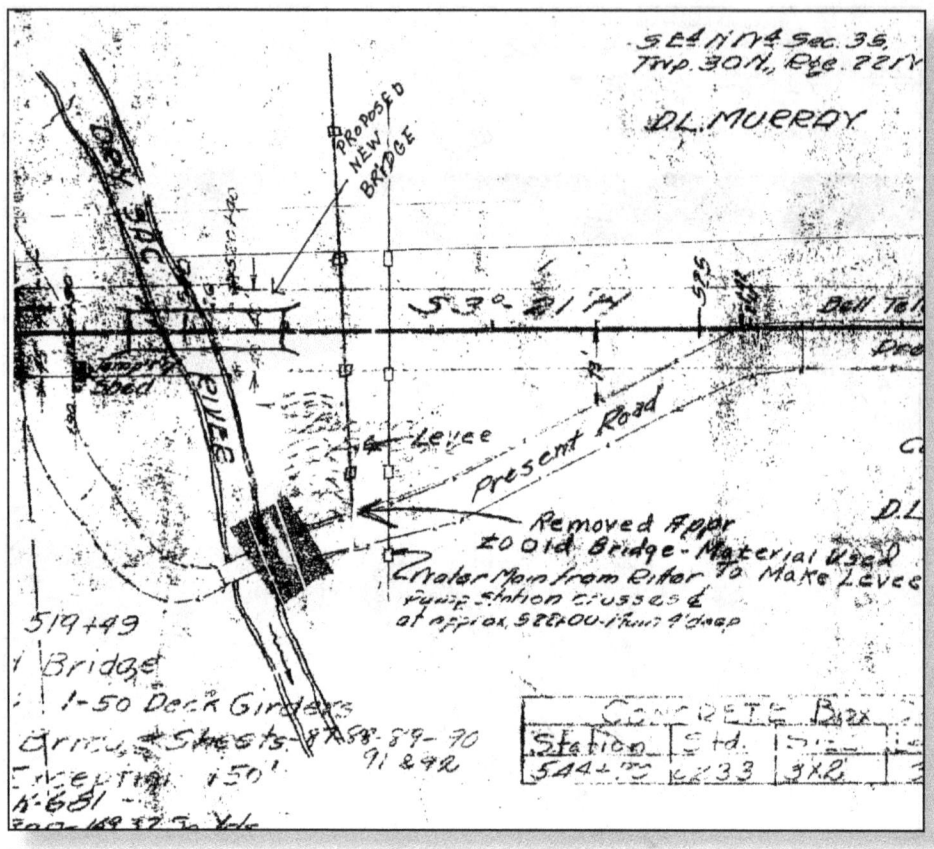

Proposed road change at Little Dry Sac River in 1928

Above diagram shows the 1928 proposed change of highway and location of the new bridge on Little Dry Sac River, Section 35. The diagram for both bridge changes and highway re-location was obtained from the Missouri Department of Transportation archives. (13) The notation "present road" is in the area where the David C. Murray Trailhead is located. See Epilogue.

Page 45 shows the 1890 bridge, a proposed new bridge and highway change on Little Sac River, Section 26. At the top of the diagram is a marker where the old bridge was located. Although that bridge is gone, an embankment approach and metal pilings are still visible.

The highway change divided the river bottom field into two fields, the land which

Early Settlers Along Boonville Road

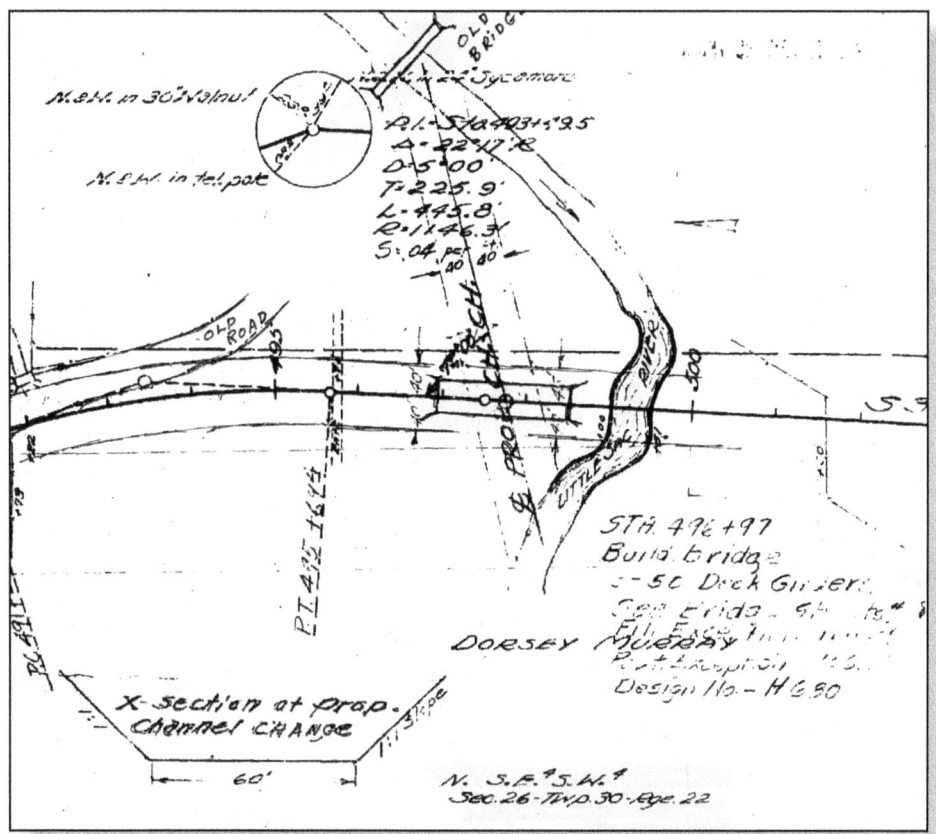

Proposed road change at Little Sac River in 1928

was noted in the early surveyor's report as Mr. Lastley's field. The surveyor was correct when he predicted the land was good for cultivation. It is still the most productive field on the Murray property.

References
(1) DeMuth I. MacDonald, *History of Pettus County, Missouri*, 1882, 406, https://archive.org.
(2) Henry C. Levens and Nathaniel M. Drake, *History of Cooper County*. (St. Louis, 1876), 174, https://archive.org.

(3) Innes and Innes, *Road Plat Book of Greene County*, (Springfield, Missouri, 1876).

(4) Greene County Court Missouri Court Minutes from 1889, Book B-89.

(5) Greene County Missouri Court Record Book M-4.

(6) Greene County Missouri Court Record Book M-28.

(7) Private Papers, Z.G. Murray.

(8) Road Petition, No. B-72, filed August 12, 1896, Greene County, Missouri.

(9) Road Petition, No. B-55, filed April 5, 1902, Greene County, Missouri.

(10) "Map of Greene County Missouri," *1876 Historical Atlas of Greene County*, (Brink McDonough & Co., 1876), at Greene County Archives, Springfield, Missouri.

(11) Greene County Court Record Book 28, December 7, 1903 term, Greene County Archives, Springfield, Missouri.

(12) Greene County Court Record Book 28, July 14, 1904, Greene County Archives, Springfield, Missouri.

(13) Diagram of road change in 1928 from Missouri Department of Transportation. Missouri Department of Transportation, Springfield, Missouri.

Chapter 4

Early Settlers in Sections 35 and 26

By 1830, the U.S. government had moved the Delaware Indians further west and Greene County, which extended from Bates and St. Clair counties in the north, Ripley County on the east, Arkansas territory on the south and Indian country on the west, was now ripe for settlement. Most of the settlers came from Tennessee and Kentucky, Virginia and North Carolina, moving on to newer and better lands. The farmers who had bragged that they "wore out a farm" in fifteen years or less were ready to try that again. There are references in agricultural books about the poor farming practices and lack of conservation that led farmers to leave their unproductive land and move on to another location. In reviewing old documents in Greene County, many of the early settlers in the 1830s had moved on to other counties by the 1850s, either showing that they, too, "wore out the land," or were looking for riches further west.

The settlers preferred wooded lands to prairies. They were accustomed to hills and they needed wood to build their houses and fence their fields. Roots of prairie grass can grow as deep as six feet, making it a laborious process to chop through the dense roots and prepare fields for planting crops.

Although few settlers could read, some might have seen publications such as Colton's *The Western Tourist or Emigrant's Guide, Mississippi Valley*, where he describes the various counties of Missouri. Colton says about Greene County: "Surface hilly, much good land with a fair proportion of prairie and timber." (1)

Settlers who came after the Civil War, such as David Murray, might have read a book such as, *Where to Emigrate and Why*, by Goddard. (2) The author described the land and commercial advantages in all the states and territories. He comments on the prairies in Missouri. "The wild pasture grasses, blue-joint, June grass, rye grass, white clover, sage and swamp grasses, furnish the natural grasses of Missouri." The pasturing season averaged about eight months. He also says, "In some seasons stock will keep fat on the pastures until December. Pasturage on the prairies is free." The free pasturage was an attraction. It was called "open range," with cattle wandering freely wherever they could find forage. However, by the time David Murray came, the prairie in Greene County was gone and farmers had divided their lands into separate fields, usually cultivating wheat, oats and corn.

Tax records, personal reminiscences and private journals indicated that there was a bustling economy among the early settlers of northern Greene County. Marriage records show that settlers frequently married a neighbor living in their section and rarely ventured very far away to find a partner.

The focus in this chapter is on the early settlers in Sections 35 and 26. These sections are where the David Murray family farm is located and where the Boonville Road passed on its way into Springfield. The patentees identified in Sections 35 and 26 are found in *An Index of the Springfield Land Office Sales Book 1833-1892*. (3)

Settlers in Section 35

Plat of Section 35. The early patentees in Section 35 included Samuel Lastley, William Fulbright, James McQuerter, Tapley Daniel, Arthur Davis, Silas Baker, James, Henry and Daniel Chandler.

Although there may have been squatters in Section 35, the first two patentees were Samuel Lastley and William (Uncle Billy) Fulbright. Both settled on the Little Sac River and Fulbright built a gristmill. They are the only two patentees who appear on the 1833 Greene County tax list.

Samuel Lastley, (also Lesley, Lessley, Lassley, Lasley, Leslie)
Samuel Lastley came to Greene County with Daniel Bird and Joseph H. Miller in 1831. He had appeared in the Maury County, Tennessee tax rolls of 1816–1828 along with the Miller brothers. They, with the Campbell and Rountree families, all left Maury County before the 1830 census. (4) Samuel and his wife, Charity Johnson, settled on Little Sac River where the Bolivar Road now crosses. (5) This is the same Mr. Lastley who is referred to in the surveyor's notes in Chapter One.

In a document named, "Greene County Mills," Ingle quotes an old-timer named Jenkins, who said when the water was low at Fulbright's mill, the millwheel would mutter: "Fulbright and Lasley, Fulbright and Lasley," meaning that it would only grind for Fulbright and Lasley, but when the water was high, the wheel would rattle away merrily, "Everybody, everybody." (6)

The 1833 Greene County Tax Assessor's list showed Samuel Lastley with 3 horses valued at $80, 1 cattle valued at $10. His tax assessment for 1834 was for 2 horses valued at $60, 1 cattle valued at $10.

In March 1835, the government surveyor began the survey of Greene County and described the land in Sections 34, 35, 26 and 27. He found some "local divergency" on the line between Sections 34 and 35, necessitating a resurvey of the line. He described the land as hilly, covered with flint stones and gravel, unfit for cultivation. It was timbered with black jack, post oak, hickory and oak undergrowth. At 32.30 chains he identified a creek 100 links wide, the left bank of said creek runs alongside of "a Mr. Lastley's field of about 35 acres." He noted the soil in that field as good for cultivation and set quarter section markers between Sections 26 and 35. (7)

The surveyor's map showed the Boonville Road crossing Section 35 with a drawing on the section lines between Sections 26 and 35, identified as "Lastley's field." Since Lastley was named in the 1833 Greene County Tax Assessment list, he apparently settled on the land and then applied for a preemption patent. This was not uncommon; many of the early settlers were "squatters" until they purchased the land.

The certificate was issued 1 September 1848. Patent #344 was for the E ½ of the NW ¼ and the NW ¼ of the NE ¼ of Section 35, Township 30, Range 22, containing 120 acres.

As noted in Chapter One, the Greene County Court records of August Term 1839 specify that Samuel Lasley was to pay all monies he had collected as overseer of the road to his successor in office.

Samuel Lastley and his wife Charity Johnson were married 11 March 1822, in Maury County, Tennessee. They are listed in the 1840 Greene County census with 1 male age 10-15, 1 male age 15-20, 1 male age 20-30, 1 male age 30-40 and 1 female age 40-50. A daughter, Martha, born in Tennessee, was married to Stephen H. Davis, 20 September 1838. (8)

In 23 January 1841, Samuel and Charity sold their land to Tapley Daniel and moved to Benton County, Missouri with Martha and Stephen. They were listed in the 1850 census of Benton County living in the Stephen H. Davis household, at which time Samuel's age was 52. The census taker listed their last name as Lesly. Government Land Office records show that Samuel Lessley purchased two tracts of land in 1845, both located in Township 38, Range 21, the same location where son-in-law Stephen H. Davis obtained patents. Davis owned several tracts of land in Benton County and was appointed postmaster at the Lessley Post Office, Benton County, in 1847.

Stephen and Martha and their thirteen children were still listed in Benton County in the 1860 census but Samuel and Charity were not included.

A public member tree on Ancestry.com said that Samuel died 20 September 1863, in Arkansas, but no documentation was provided. It is possible that the Samuel who died in 1863 was confused with the Samuel Lastley who lived in Benton County. A probate file in Cedar County, Missouri described a lawsuit against Samuel Lastley's (Lesley) estate because he unlawfully took two horses from his neighbor in August 1862. This could have occurred as Confederate and Union troops waged war across the Ozarks. The situation is not explained, nor are the circumstances surrounding Lastley's death. The plaintiff was Adam Eslinger, a U.S. soldier. The jury awarded $250 damages to Eslinger to be paid from Samuel G. Lesley's estate in 1865. (9) Lesley's widow was Mary, with Louisa and George named as minor heirs. Census records spell Samuel's name as Lesley and there are two other entries for a Lesley family living in Cedar County in 1860. It is not known if this Samuel is related to Samuel Lastley. In the 1870 census of Benton County, Charity Leslie, age 74, was head of the household with grandchildren, Napoleon and Josephine (Davis) living with her. In the 1880 census of Benton County, Charity Lasley is living in the household of her daughter, Martha J. Davis. Martha's husband, Stephen, died 31 December 1879, in Benton County. Charity died 30 October 1887 in Benton County, according to the informant on Ancestry.com and Martha died 1901, Benton County, Missouri. Martha and Stephen are buried in the Davis Cemetery, Cross Timbers, Benton County, Missouri. I have found no information about the burial sites of Samuel and Charity.

William Fulbright
The history of the Fulbright family is well known in Greene County and will not be covered here. He is named because of his relationships with families in the neighborhood. He obtained several land patents; the one of interest is the E ½ of the SE ¼ of Section 35, Township 30, Range 22, containing 80 acres. His gristmill was located on Sac River and is the one referred to in the earlier story about Samuel Lastley. He had no relationship with the Murray family because they moved to Greene County after he died. However, early neighbors often met at his mill. See the story of the stabbing of C.B. Owens under The Banfield Family in Chapter Five.

His probate file located in the Greene County archives did not list any relationship with settlers in Sections 35 and 26. William Fulbright died in 1843 and there were later references to sales and purchases in Greene County by his wife Ruthie.

Following are names of other settlers who came to Section 35 a few years after Lastley and Fulbright.

Arthur Davis
Arthur Davis was born in North Carolina and was living in Greene County

by 1836 because he is listed as serving on juries in 1836, 1837, 1839 and 1847. He was listed in the 1840 Greene County census with 2 males age 20-30, 1 male age 40-50, 2 females age 15-20 and 2 females age 40-50. He obtained a patent for the NW ¼ of the NW ¼ Section 35, Township 30, Range 22 issued on 10 April 1843. This parcel was sold to Tapley Daniel on 17 January 1850, comprising part of the 320 acres that Daniel sold to David Murray in 1867. The original patent was given to Tapley Daniel, who gave it to David Murray.

When Arthur Davis sold his property to Tapley Daniel in 1850, he also sold the NE ¼ of Section 24, Township 30, Range 22, reserving 50 feet for a graveyard. Since he reserved the area, some of his family members must have been buried there. That graveyard is now called the Wilson graveyard and is located on the property of the Springfield Northwest Wastewater Treatment Plant, just west of Highway 13. This property was acquired by Joseph B. White, son-in-law of Tapley Daniel, and sold to William Wilson on 22 March 1858. (10) Nothing further has been found about Arthur Davis in Greene County.

Silas Baker
Silas Baker was an early educator in the northern section of Greene County. His relationship in this history is only as a purchaser of the SW ¼ of the NE ¼ of Section 35, Township 30, Range 22. There is no evidence that he lived in this section.

He died 23 November 1840 in Greene County. His probate file, #443, Greene County, indicated assets of two sections of land, and personal belongings relating to education, consistent with employment as a teacher. He died leaving no heirs. Marsha Rising has a history of Silas Baker in her book. (11) Silas was buried in Old Salem Cemetery close to the Wallis family. Jeptha Wallis was executor of his estate and the property was sold to Bryant Nowlin. It is possible that Baker bought these properties as investments and never lived on the property. It was common for teachers to board with families and his probate file showed that several families owed him money for tuition.

James S. McQuerter (McQuirter, Mcqrtr)
James McQuerter is mentioned here because he was a patentee in Section 35.

He had no direct relationship with the Murray family, but he sold land to Qualls and John Banfield.

A public member tree on Ancestry.com stated that James Simerall McQuerter was born 28 December 1811 in Tennessee and died 25 December 1891, in Greene County, Missouri, buried in Maple Park Cemetery in Springfield, Missouri. Ozarks Genealogical Society researchers identifying burials at Maple Park found that James S. McQuerter was buried there, but they stated there was no stone. (12) Contact with the administrator at the cemetery office indicated that James McQuerter's stone is located on the James A. Dameron lot.

In addition to the stone showing birth and death dates for James Simerall McQuerter, there are three other McQuerter stones on the Dameron lot. Grace McQuerter, born 22 December 1884, died 28 January 1904. William McQuerter, born 10 March 1847, died 29 May 1899. (He was listed in the 1850 Greene County census as the son of James S. McQuerter). Malvina McQuerter, born 17 October 1846, died 17 May 1898. The stone for Grace was placed in the cemetery 30 January 1904 and the stones for James, William and Malvina were placed there on 5 March 1904 indicating that those three individuals had either been previously buried elsewhere or that these stones were placed in memory of the individuals.

Grace was listed in Webster County birth records with William as her father, age 40, living in Mansfield, Missouri, but there was no name or birthplace listed for Grace's mother. (13) All information pertaining to her was missing, except that she was age 35. In the 1900 census, Grace is living in the home of James A. Duncan and his wife Mary, on Jefferson Street in Springfield. Grace is identified as niece, age 14. The name Duncan rather than Dameron must have been an error on the part of the census taker as well as Mary's date of birth. I believe the 1880 census information is correct because Mary Frances Dameron's tombstone states that she died in 1924 at age 69.

In the 1880 census of Greene County Missouri, James A. Dameron, age 26 was listed as an agent for a lightning rod company, his wife Mollie age 26 and his sister-in-law Laura McQuirter age 21 were living on Boonville Street. This explains why the McQuerters are buried on the Dameron cemetery plot. James Dameron's wife was Mary Frances, (Mary, Mollie, M.F.) child of James S. McQuerter, as was Laura.

Both Laura and M.F. appeared on the 1860 Greene County census as children of James S. and Jane McQuerter.

James S. McQuerter first appeared in Greene County in the 1850 census of Campbell Township. He stated his age as 39, born Tennessee with a real estate value of $700. Family members were Jane, age 36, born Tennessee, John and Wesley ages 17, Nancy age 16, Lucinda age 14, Louisa age 12, P.M. age 6 and Wm. G. age 4, all born in Missouri. Because the children were all born in Missouri, this could be the same James McQuerter who purchased 40 acres by land patent in Moniteau County, Missouri: Section 10, Township 45, Range 17W, 14 November 1835.

There was a James S. McQuerter in the 1840 census of Willow Fork, Morgan County, with 1 male age 5-10, 1 male 20-30, 3 females under 5 and 1 female 20-30.

In Greene County, James S. McQuerter entered a preemption claim for the NW ¼ of the SE ¼ and the NE ¼ of the SW ¼ Section 35, Township 30, Range 22. Patent #15402 was issued 10 March 1856. Since McQuerter was enumerated in the 1850 and 1860 census living in Campbell Township, not Robberson Township, it is unusual that his patent was a preemption patent. It may have been that he was filing for a first right of purchase, rather than claiming ownership by settling on and developing the land.

On the 1851 tax assessment list in Greene County, James McQuerter had 2 horses valued $90, 4 cattle valued $20, money on notes $50, Section 13, Township 30, Range 22, 80 acres valued at $200. In reviewing the original tax book, it was noted that the recorder placed a "ditto" under the section number, shown in a previous line. Because the patent was not issued until 1856, and James notes Section 35, Township 30, Range 22 in the 1856 tax assessment list, it is felt that the Section noted on the 1851 tax assessment list is in error.

On the 1856 tax assessment list, James S. McQuerter had 3 slaves, valued at $950, 7 horses valued at $500, 4 cattle valued at $40, and 1 timepiece valued at $3. Real estate listed was Section 35, Township 30, Range 22, 80 acres and Section 35, Township 30, Range 22, 40 acres. Total value was $960.

James McQuerter sold the above properties located in Section 35 in 1857. He sold the NW ¼ of the SE ¼ and NE ¼ of the SW ¼ consisting of 80 acres for $700 to Qualls Banfield on 1 June 1857. (14)

James McQuerter and Jane sold to John Banfield for $600, 20 acres m/l of the SW ¼ of the NE ¼, Section 35, Township 30, Range 22, on 14 December 1857. (15) On 20 December 1857, they also sold 20 acres m/l of the SW ¼ of the NE ¼ of the NE ¼, Section 35, Township 30, Range 22 to Qualls Banfield. (16) These two sales are the same 40 acres patented by Silas Baker.

According to the 1870 census, James McQuerter and his wife were living in Webster County, Missouri. Jane had died, date not known. James S. McQuerter married Sarah Durham in Webster County 11 March 1869.

Tapley Daniel (Daniels)
Tapley Daniel had a close relationship with the Murray family. The land he purchased was sold to David Murray when Daniel and his family moved to Miller County, Missouri.

Tapley Daniel stated on census reports that he was born in North Carolina. Several family tree postings on Ancestry.com state that his full name was Tapley Passley Daniel. There is a Taplay Daniel listed in the 1830 census of Sumner County, Tennessee: 1 male under 5, 1 male age 20-30, 1 female 15-20 and 1 female 20-30. No marriage record has been found in Sumner County. Tapley's wife was Keziah Thornhill. There is a marriage bond in Sumner County between Thomas Daniel and Judy (Juditte) Thornhill, 10 April 1816. In the Scholastic Population census of 1838, Sumner County, Tennessee, Tapley Daniel lived in District #6 with 1 child in school; Thomas Daniel lived in District #5 with 4 children in school. Although the relationship between Tapley and Thomas and Keziah and Judy is not known, there were enough interactions between the families to suspect brothers married sisters.

29 December 1846, Tapley loaned $300 to Thomas and Judy to buy NE ¼ of SW ¼ Section 4, Township 28, Range 22. (17) Thomas and Judy are listed on the 1850 Greene County census: Thomas age 60, Judith age 56, daughters Marguerette age 30, Nancy age 28 and Sarah age 16. However Margaret (Marguerette) and Nancy are already married, perhaps they were in the house caring for their parents.

Margaret Daniel married John M. Crockett, 28 May 1843 and Nancy Daniel married Joseph Cain, 4 March 1841. (18) In the 1850 census of Greene County, Missouri, the family living next to Thomas Daniel is Joseph Kain (Cain) age 30, his wife Nancy age 28, children William L. age 3 and Letticia age 1. Also in the 1850 census of Greene County, Porter Township, there is John M. Crockett age 32, Margarette age 30 and their children, Andrew age 6, Thomas age 4, Martha age 3 and Mary age 1. It appears that Nancy and Margarette (Margaret) were enumerated twice.

John Goodridge married Sarah Daniel, 26 December 1854. (19) The relationship of Thomas Daniel's children is further verified in a deed whereby the heirs sold their rights to Josiah (Joseph) Cain. On 15 February 1855, William S. Daniels, John and Margaret Crockett, John and Sarah Goodrich, Andrew and Rachel Ricketts (heirs of Thomas) sell to Josiah Cain for $10 each paid by Cain, Section 4, Township 28, Range 22, consisting of 40 acres. (20) It is not known if Nancy Cain (wife of Joseph or Josiah) is still living.

Thomas Daniel made his final will 21 June 1851. The appraisers of his estate, Cannefax and Moore, state that he died 4 September 1851. He did not name his children in the will. The children, named in the above deed, should not be confused with the children of Tapley and Keziah.

On 23 January 1841, Samuel Lastley sold his 120 acres to Tapley Daniel and soon after that Daniel began acquiring land in Greene County. In the Greene County tax assessment book of 1843, Tapley is assessed for 3 horses valued at $100, 2 cattle valued at $14, real estate located in S35, T30, R22, 120 acres valued at $720. This is the land that Lastley acquired by patent.

On 3 June 1848, Tapley received land patent #8051 for the NW ¼ of the SE ¼ of Section 26, Township 30, Range 22 containing 40 acres. 3 April 1848, he received patent #8050 for the SW ½ of the SE ¼ of the SW ¼, Section 26, Township 30, Range 22, containing 80 acres. On 1 January 1849, land patent #9105 was issued for the SW ¼ of the NE ¼ in Section 26, Township 30, Range 22, containing 40 acres. His final land patent was #18.110 for the SW ¼ of the NW ¼ of Section 35, Township 30, Range 22, containing 40 acres, issued 15 May 1857.

On 17 January 1850, Tapley Daniel purchased the 40 acres patented by Arthur Davis, land that Davis obtained by patent #1239 on 10 April 1843. This completed Daniel's acquisition of 320 acres in Sections 26 and 35.

In 1851 Tapley Daniel was assessed for 3 horses valued at $75, 5 cattle valued at $50 and real estate located in S35, T30, R22 valued at $360. Fortunes had increased for Tapley by 1856 when he was assessed for 3 horses valued at $135, 6 cattle valued at $75, 1 timepiece valued at $3, and real estate located in S35, T30, R22 80 acres; S26, T30, R22, 80 acres; S26, T30, R22, 40 acres; S24, T29, R22, 2 acres; S35, T30, R22, 120 acres.

On 14 June 1852, Tapley purchased the land, buildings and machinery from the Eli Jessup estate located Section not listed, T24, R22 in the city of Springfield. The tannery was the first in Springfield, operated by Thomas Jessup, who came to Springfield in the early 1830s. It was operated in the Valley of the Jordan, west of the southwest corner of Boonville Avenue and Mill Street. Thomas' son, Eli, died in 1848 and the real estate and tools were appraised at $566. Tapley bought the land and stock for $200, which he sold to Charles Stark, 11 February 1857, for $475. (21)

In the 1850 census of Greene County, Tapley Daniel and his wife were living in Robberson Township, household #536. He stated his age as 65, Kiziah was 55, with son John age 18 and daughter Lucinda age 12. Household #540 was that of Joseph White age 24, Martha age 20. She was the daughter of Tapley and Kiziah and they were living next door to Joseph's father, A.D. White. Tapley and Kiziah sold to Joseph White for $400 and for the natural affection to my daughter Martha White the NE ¼ of Section 34, Township 30, Range 22 and the SW ¼ of the SE ¼ of Section 27, Township 30, Range 22 in Greene County Missouri for a total of 200 acres, 6 December 1852. (22) This confirms the relationship between Tapley and Martha.

In the Circuit Court of Greene County, 13 March 1858, John Daniels, Robert Cannon, John Dickens and Hiram Thornhill were accused of unlawfully betting "a large sum of money to wit, the sum of eighteen dollars and fifty cents upon a game of chance played at and by means of a certain gambling device called cards, adapted devised and disguised for the purpose of playing games of chance for money and property." Bondsman for John was his father, Tapley. Bondsman for Hiram Thornhill was J.B. White, son-in-law of Tapley, and possibly, a cousin by marriage to Hiram. Hiram was the son of Epperson Thornhill, who may have been related to Kiziah Thornhill Daniel and Judith Thornhill Daniel. Stephen Dickens was bondsman for his son, John. The case was ultimately dismissed, no charges.

In the 1860 census of Robberson Township, Greene County, Missouri, Tapley Daniel is listed as age 60, Kizzy is age 60 and Lucinda age 20, value of real estate is $3000. The next family listed is John Daniel age 35, wife Drucilla age 34, Willis age 6, Tapley age 5 and Wiley 3 months. John Daniel's occupation is listed as carpenter. John married Drucilla (on license listed as Druciller) Halburt or Hurlbert, 24 February 1853. (23)

In July 1865, Roger Q. Banfield sued John Daniel for wrongfully taking his horse valued at $135 and clothing valued at $40. (24) John Daniel was not a resident of Greene County. Later, the court ordered the sale of two mares and a filly (allegedly owned by Daniel) with proceeds to go to Banfield. (25) In August 1866, the Circuit Court stated that Banfield was the injured party and the money received by the sheriff from sale of the animals was to go to Banfield. John Daniel was the son of Tapley Daniel, both of whom were neighbors to Banfield. It is possible that Daniel took the horse and clothing and went south during the Civil War. In his testimony, John stated that he left the state in February 1862. This was the time when a number of southern supporters followed General Sterling Price to Arkansas. Banfield and his family were Union supporters. More on this lengthy case and the final judgment will be found in Chapter Six, David Murray Comes to Missouri.

On 5 September 1867, Tapley and Kiziah Daniel sold 320 acres to David C. Murray of Wyandot County, Ohio, and apparently moved with their daughter Lucinda and husband Giles Williams to Miller County, Missouri. The sale was for $6000 with a mortgage of $3500 payable in seven annual installments at six percent interest. Daniel and Murray apparently corresponded with each other, as there are several letters in David Murray's papers signed "Your friend, Tapley Daniel." His final letter, December 1872, says that he is blind, and wishes he could see his old friends in the neighborhood again. He expresses sorrow at the death of his old friend.

Early Settlers Along Boonville Road

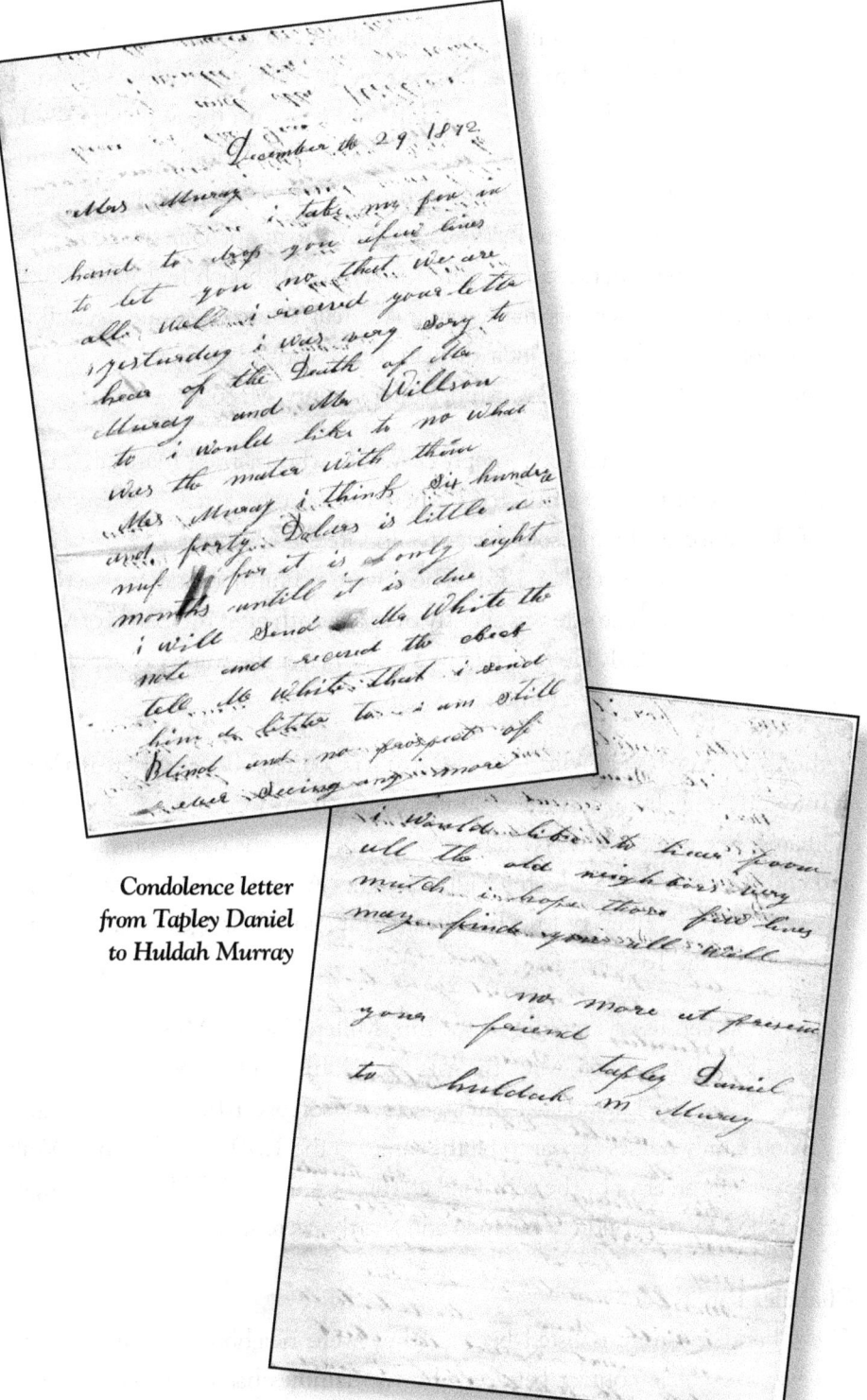

Condolence letter from Tapley Daniel to Huldah Murray

On the 1870 census of Franklin Township, Miller County, Missouri, G. Williams, farmer, age 21 is listed with his wife, Lucinda age 30 and their daughter Victoria age 1. The dwelling next door has a T.P. Daniel residing with the William Davidson family. He is age 90, a retired merchant, born in North Carolina. I believe this is Tapley because of the initials and also because of his close proximity to Lucinda. His age has varied in each census, but if we believe his age of 65 in the 1850 census, he could be in the nineties category in this census. Although he earlier listed his occupation as a farmer, his many profitable real estate investments indicated an entrepreneurial bent. Lucinda died in 1922, verified by her Missouri death certificate #40748 in Saline Township, Miller County, Missouri.

According to Goodspeed (26) Joseph B. White who married Martha C. Daniel lived in Greene County until 1862 when he went to Texas and entered the Confederate forces. Joseph sold 320 acres in Greene County to Zachariah Sims, a neighbor, on 26 September 1861. There were no further land transactions by him after that date. The sale was slightly over a month after the Battle of Wilson's Creek, 10 August 1861. He returned to Missouri after the war was over and lived there until 1883 when he returned to Arkansas.

In the 1870 census of Franklin Township, Miller County, Missouri, there is a Walter White and family living just three families distant from Tapley Daniel and Lucinda Williams. The name Walter is likely incorrect, because the names of wife, Martha, and children, Sophronia, Hugh L., Julia E. James A. and Caladonia, are the same as those family members of Joseph Bolivar White and their ages would be correct in relation to the 1850 census.

In the 1880 census of Franklin Township, Miller County, Missouri, family #32 listed a J.B. White, age 55, M.C. his wife age 50, his son H.L. age 21, J.E. son age 19, J.A. son age 15, Lucinda daughter age 13, M. son age 10 and M.C. son age 10. Ages and family names appear to be the same, so the 1870 census listing a Walter White is likely an error by the census taker. Family #31 is J.G. Williams age 30, wife Lucinda age 35 and children. Lucinda and Martha were sisters.

Chandler Family
The Chandler family is listed because they were neighbors to David Murray; however very little contact between the two families has been found. Stephen

Chandler's daughter married Willis Dickens and the Chandler family had contact with William Jones who was buried in the Murray Cemetery.

Daniel Chandler took out a patent for the SE ¼ of the SW ¼ of Section 35, Township 30, Range 22. The patent #29.018 was issued 1 October 1860. He and his family must have moved to Greene County between 1840 and 1850. He is listed in Halifax, Virginia in the 1840 census with two males under 5, 1 male age 5-9, 1 male age 20-39, and 1 female age 20-29. He and Sarah M. Hall had a marriage bond issued in Person County, North Carolina on 25 November 1833.

Daniel Chandler and family were listed in the 1850 census of Greene County, Missouri in Robberson Township. Daniel was age 41, farmer, born Virginia, Sarah M. was age 40, born Virginia, William P. age 14, James M. age 13, John age 11, Margarette age 9, Henry age 7, Elizabeth age 6, all born in Virginia, Willis age 4, Nancy age 4 months, both born in Missouri. Since Willis was born in Missouri, this would indicate a move sometime around 1845.

In the 1860 census of Greene County, Robberson Township, Daniel was listed as age 51, farmer, real estate value $2500, personal property value $1000, born Virginia. His wife was Sallie age 43, born Virginia, John age 20, Margaret age 26, Henry age 15, all born in Virginia, Elizabeth age 14, Willis age 12, Mariah age 8, all born in Missouri.

A public member tree on Ancestry.com states that Daniel married Sarah Jane Marley in 1865 in Greene County, Missouri; however the marriage has not been found in Greene County records. In the 1870 census of Greene County, Daniel was living in Campbell Township, with his occupation listed as tenement landlord. His age was 61, born Virginia, Sarah, his wife was 37, born North Carolina, Orin age 16, worked as farm hand, Lydia age 14, Franklin age 12, Sarah age 4 and Caroline age 11 months, all children born in Missouri. If Daniel and Sarah were married in 1865, it appears that the older children may be Sarah's children from a previous marriage.

In the 1860 census of Robberson Township, Stephen Chandler lives just three houses away from Daniel Chandler. Steven is identified as the son of Daniel and Sarah Chandler in the history of *The 24th Missouri Volunteer Infantry* (27). Stephen enlisted in Company A on 28 February 1862 in Springfield.

Missouri records listed a marriage between Stephen Chandler and Sarah Jane Hall, 6 April 1866, marriage performed by John M. Chandler, minister of the gospel. (28) This means that the marriage was performed by Stephen's brother, John M. Chandler, who will be mentioned later in connection with other neighborhood marriages.

The 1870 census of Robberson Township listed the neighbors of the Murray family. Stephen Chandler was in household #97. He was listed as 35 years of age, farmer, personal property $300, born Virginia. His wife Sarah J. was 21 years, born Missouri. Children were Martha J. age 2, John age 1, Maude 4 months, all born in Missouri. Household #95 was Qualls Banfield, household #96 was Ann Warren (probably Lucy Ann Warren Banfield's mother), household #98 was Henry Chandler, household #99 was William Jones (buried in the Murray Cemetery), and household #101 was David Murray.

In the 1880 census of Greene County, Robberson Township, Stephen Chandler was listed as 43 years, his wife Sarah J. 31 years with children: Martha age 13, John age 12, Eliza E. age 10, George age 8, Viola age 5, Jesse age 2 and Henry 3 months. Viola later married Willis Dickens, son of Stephen Dickens and Judy Scott Dickens. See Dickens family history in this chapter.

In the 1900 census, Stephen and Sara were living in Taney County, Missouri. In 1910, Stephen, widowed, was living in Ward 7, Springfield, Missouri, listed his occupation as "own income."

James Chandler, son of Daniel, entered a claim for land in the NW ¼ of the SW ¼ of Section 35, Township 30, Range 22. The patent was a preemption patent issued 27 March 1861. James Chandler married Judy Akins 18 March 1858 in a ceremony performed by C.C. Williamson, a local Methodist minister. James H. Chandler was listed in the 1890 Surviving Schedule of Veterans of Missouri, as serving for 10 months in Company A, 24th Missouri Infantry, enlisting 5 August 1861.

Henry R. Chandler, another son of Daniel, entered a claim for land in the SW ¼ of the SW ¼ of Section 35, Township 30, Range 22. The patent was issued 25 March 1872. He married Eliza McKey 28 July 1863, marriage performed by J. Spain, minister of the gospel. This is the same J. Spain who married Anna Elizabeth Austin. See history of the Austin family.

Henry died intestate in 1890, leaving a small amount of personal property to his widow, Eliza.

Settlers in Section 26

The Government Land Office records show Tapley Daniel as the first patentee in Section 26. Certificates #9105, #8050, #8051 granted him 160 acres in this section. His purchases were detailed earlier. Tapley Daniel sold 40 acres (patent #9105) to Samuel Austin in 1853. Because Austin was an early setter, the Austin family is detailed in this section. David Murray and Thomas Wilson are not included here because they acquired their properties much later. Note also the sections which are dedicated to the railroad.

Plat of Section 26. The early settlers in Section 26 were Tapley Daniel and Samuel Austin. (3)

Samuel Austin III

The Austin family came to Greene County in 1835. Most of the children settled in other parts of Greene County; John and Samuel settled in Section 26 in the 1850s. They had several interactions with the Murray family. To understand these, it is necessary to state the history as it has been found to date.

Information on the family was first found in "Genealogies of Early Springfield Families." (29) The information was supplied by descendants of Samuel Austin II (sometimes referred to as Samuel Austin, Jr.) The children of Samuel Austin II and Catharine Payne, all born in North Carolina were: Mary, Anna (deaf mute), John born 1804, Greene born 1805, Jennie, Catharine born 1809, Sarah born 1811, and Samuel III born 3 February 1814. The notation that Anna was deaf and later census notations raise the question of a possible hereditary deafness in the family. In the 1850 census, Samuel Austin age 82 and Samuel Austin age 45 are both listed as deaf. Daughters Catharine and Sarah are noted as being deaf or unable to write in various census reports. John, Samuel III, Catharine and Sarah all have roles in this narrative.

John Austin received two patents in Greene County. They were #1479 and #1480, but they were not for property in Section 26, Township 30, Range 22. He later owned property in Section 26, sold to him by his brother Samuel. John first married Rachel Freeman in North Carolina, and after her death married Louisa Williams in 1848 in Greene County, Missouri.

In Missouri records there is a marriage recorded for Samuel Austin (son of Samuel Austin II) and Elizabeth Jane Calvert, 28 December 1845. (30) It is more probable that her name was Halbert because the Calverts listed in the 1850 Greene County census were living on the east side of the county closer to Webster County. However, just eight residences away from Samuel and Elizabeth was the Jesse Halbert/Halburt family. There were other Halberts who married neighbors in Robberson Township. John Daniel, son of Tapley, was married to Drucilla Halbert. It is probable that Drucilla and Elizabeth Jane were sisters.

In the 1841 tax records of Greene County, Samuel Austin is assessed for two cattle valued $15, Samuel Austin Jr. is assessed for 1 horse valued $30. Sam Austin Jr. is probably the same as Samuel Austin III. John Austin is assessed for 1 horse valued $30, 4 cattle valued $40.

In the 1851 Greene County tax records, Samuel Austin, Sr. is assessed for 1 horse valued $25, 2 cattle valued $10. Samuel Austin, Jr. is assessed for 1 horse valued $25, 4 cattle valued $20. In the 1856 Greene County tax records, Samuel Austin is assessed for 4 horses valued $160, 2 cattle valued $20, 1 time piece valued $3 and

40 acres located Section 23, Township 30, Range 21. This is a puzzle because this property was acquired by John Austin, patent #1479, 10 April 1843.

In the 1850 Greene County census of Robberson Township, family #535 was Samuel Austin, age 35, with real estate value $200, born North Carolina, listed as deaf. Elizabeth, his wife, was age 24, born North Carolina. Children were John C. age 4, Sarah A. M. age 3 and Elizabeth age 1. A child named Simon Breach, age 9 was living with them.

Family #536 was Tapley Daniel and family #544 was Jesse Halburt.

In the 1860 census of Greene County, Robberson Township, Sam Austin was age 40, occupation farmer and teamster, $150 personal property. Elizabeth must have died because living with him was Cate age 45, born Tennessee, Calvin age 15, Sarah age 12 and Elizabeth age 10. It seems likely that Cate was Samuel's sister, Catharine. In the family history, she was born in 1809 and he was born in 1814, which would correspond with the ages on the census.

On 17 December 1853, Tapley Daniel sold to Samuel Austin, the SW ¼ of the NE ¼ of Section 26, Township 30, Range 22, consisting of 40 acres. (31) On 11 June 1856, Samuel borrowed $45 from Joseph Gott secured by a mortgage on the above property. (32) Possibly to repay the debt on the property, Samuel sold the land to his brother, John Austin 10 September 1857. (33)

In the early 1800s the U.S. government dedicated lands for the railroads planned across the country. The gifts of public lands were made in exchange for the railroad laying tracts in certain areas. As land values increased, the railroads sold these lots and used the proceeds to pay for materials and labor to continue their expansion.

In Greene County, the South Pacific Railroad began releasing tracts that it would not use for construction of future rail lines. On 24 September 1870, the South Pacific Railroad Company sold 120 acres for $300 to Samuel Austin located on the E ½ of the NE ¼ and the NW ¼ of the NE ¼, Section 26, Township 30, Range 22. (34)

In the 1860 census sisters Cate and Sally were living with their father, Samuel. The census taker made no note of anyone deaf in the family; he only noted that

Cate was unable to write and Sally was unable to read and write.

In the 1870 census, Robberson Township, family #67 was Samuel Austin, age 56, farmer, born North Carolina, John C. age 21, Sarah M. age 19, Jane age 17. Family #66 was that of Hosea Mullings. Living with the Mullings family was Catharine Austin, age 60, born North Carolina, noted "cannot write."

On 14 November 1871, Samuel Austin and David C. Murray entered into an agreement to purchase from the South Pacific Railroad Company the N ½ of the NW ¼ of the SE ¼ of Section 26, Township 30, Range 22. (35) David Murray died in April 1872, and his widow married Qualls Banfield. On 8 August 1874, Huldah M. Banfield and R. L. (Qualls) Banfield gave a quitclaim deed to Samuel Austin. (36) See Chapter Six.

On 9 July 1877 John Austin and his wife, Eliza A, sold to Elizabeth Jane Austin the SW ¼ of the NE ¼ of Section 26, Township 30, Range 22, reserving 40 square feet, including the grave of John Calvin Austin. (37)

This sale presents a puzzle. Why would he sell the land to his daughter and reserve the gravesite of John C.? Was John his son?

The first clue to answering the puzzle was in looking at the census reports. In the 1850 Greene County census, Samuel Austin had a daughter named Elizabeth age 1 year and John Austin had a daughter Elizabeth, also 1 year. Sam's daughter Elizabeth was listed by the name Elizabeth again in the 1860 census, but in the 1870 census, she was listed as Jane. John's daughter, listed as Elizabeth in the 1850 census, was listed as Ann in the 1860 census.

A little more sleuthing at the Greene County Archives revealed an answer in the Greene County Missouri Deed Book where an unusual entry was found. E.J. White (Elizabeth Jane) says that to the best of her knowledge and belief that James M. Dickens and Sarah M. Austin were married 7 March 1872, by Rev. J. Chandler. Next, Sarah M. Dickens says that to the best of her knowledge and belief that G.W. White and E.J. Austin (Elizabeth Jane) were married 19 July 1877 by the Rev. J.C. Chandler. The two sisters were swearing each other's marriage dates. The acknowledgement was sworn before Justice of the Peace George B. Rabe on 13 February 1893. (38)

So the question arises, which Elizabeth bought the property from John Austin for $400? It seems more plausible that Elizabeth Jane Austin, daughter of Samuel, purchased the land on 9 July 1877, and then married George W. White on 19 July 1877. George White is recorded paying taxes on this piece of land in 1878–1879.

Elizabeth Ann Austin (daughter of John) married Rev. James C. Spain, who was the moderator of the Cedar Bluff Baptist Church in Fair Grove, Missouri. In another Greene County, Missouri deed book entry, there is a record of marriage between James Spain and Anna Elizabeth Austin on 20 May 1874. The marriage occurred at the house of James M. Dickens and was performed by Rev. John Chandler. The marriage was recorded 19 November 1892, witnessed by Louis H. Dickens, who said the marriage had never been previously recorded. (39) This Elizabeth could not have been the Elizabeth Austin in the 1877 sale because she was already married to Rev. Spain. Rev. Spain's divorce from Sophronia Austin was granted 19 May 1874, and he was married to Elizabeth Ann the following day. Rev. Spain was granted the divorce as the injured party. Perhaps because of the divorce and quick remarriage, the marriage was not reported earlier.

In the 1880 census of Barry County, Missouri, James Spain age 45, farmer, was listed with Anna E. his wife, age 30, William W. age 22, Robert S. age 19, Jesse age 17, Mary J. age 15, George age 14, John age 7, Claude age 4, and Albert age 3. William, Robert, Jesse, Mary J. and George were children by the first wife, Sophronia.

On 6 October 1939 Elizabeth Ann Spain died in Monett, Barry County, Missouri. The death certificate #36600 gave her date of birth as 11 October 1849. The informant only knew that her father's name was Austin and her mother's name was Williams, which would confirm the information on the Austin family found in Marsha Rising's book. (40)

The other puzzling and unanswered detail about John Austin's sale of land to Elizabeth Jane is reserving the grave of John Calvin. At first glance, this would indicate that John C. is John's son; however, John C. is enumerated in the 1850, 1860 and 1870 census living in the household of Samuel Austin. The reservation of land could be out of respect for a brother's son, but in that case it would be strange that Samuel did not bury John Calvin on his own property, which was adjacent to John's property.

On 18 September 1884, Samuel Austin sold for $600 the SE ¼ and the W ½ of NE ¼ Section 26, Township 30, Range 22, to Sarah M. Dickens, reserving the right of premises and the use of the farm and premises during his lifetime. (41) Nothing further has been learned about Samuel after 1884. Was he buried in one of the unmarked graves in the Murray family cemetery?

To complete the information on the Austin family, there is more to be learned about the sisters, Catharine and Sarah. In the 1880 census of Greene County, John Austain age 80, farmer, was living in Franklin Township, just east of Robberson Township with his wife, Eliza age 51, Margaret A. age 22, daughter, Francis M. age 19, son, Catherine, age 60, sister, and Claudy, age 1, grandson. Claudy is Margaret's son, born 11 January 1879. Margaret died 20 January 1940, verified by Missouri death certificate.

Catharine Austin, listed above, was living with a different family in each census report. In the 1850 census, she was living with her father, Samuel and was identified as being deaf and unable to read and write. In the 1870 census she was living with the family of Hosea Mullings, identified as unable to write. In the 1880 census she was living with her brother John. The *Springfield Leader Democrat*, 23 March 1900 reported "Mrs. Katherine Austin, died yesterday at the home of J.M. Dickens, six miles north of the city. She was buried in the Murray Cemetery." There is no stone for her; she lies in an unmarked grave. James M. Dickens was married to Sarah Austin, Catharine's niece. See the section on the Dickens family.

The other sister, Sarah, also had a history of living in different homes. She was living with her father, Samuel II, in the 1850 census, with the note that she was unable to read and write. The census taker had written "deaf" beside Samuel and Catharine. Did he move the remarks up a line? Should "deaf" have been written beside both Sarah and Catherine? In the 1860 census, Sally Austin, age 45, was living with James McMurry. Her occupation was sewing and spinning and she was identified as a person who cannot read or write, no note of deafness.

In the 1870 census of Greene County, Sarah Austin, age 50, seamstress, was living with the family of William Chrisman, Campbell Township. The census taker did not indicate an occupation, but noted she was deaf.

In the 1880 census of Greene County, there was a Sarah Austin, age 60, widow, born in Tennessee, listed as residing at the Alms House, (local institution for the poor and disabled) located in Campbell Township on the Hartville Road. She was identified as insane and deaf and dumb.

In the *History of the Greene County Alms House* there was an entry for Sallie Austin, age 84, admitted 11 January 1879. She died 23 December 1890. (42) It is not known if this is the same Sarah (Sallie) Austin as listed in the 1880 census. The age listed in the census report would be more accurate than the one listed in the Alms House Book. Her birth date was 12 March 1811. She was listed as insane in both the census and the Alms House Book. However, in the nineteenth century many deaf people were placed in mental institutions. When they could not communicate verbally, and they could not read and write, how would the deaf person know if the interviewer guessed correctly? No information regarding her burial has been found.

Dickens Family
The Dickens family settled in Section 25, rather than Section 26, and they had no marriages with the Murray family; however the two families were neighbors and several Dickens were buried in the Murray Cemetery. There were intermarriages between the Dickens, Austin and Banfield families.

There is a post on Ancestry.com under the heading, Dickens Family Tree, which says that Jesse Dickens and his wife, Polly McDerment, were parents of sons: Stephen born 1807, Daniel born 1814, William B. born 1814, Joseph H. born 1816, and Joshua born 1820. There is no indication that Daniel and William are twins, or whether there is an error on their birthdates. According to the Ancestry entry, Daniel, Joseph and Joshua all died in Tennessee. Stephen and William B. came to Missouri. There is no mention of a daughter who was discovered when researching Solomon Cotner. See Chapter Seven, Burials in Murray Cemetery.

Stephen, the oldest brother, was in the 1850 census of Campbell Township. He was age 45, wife Nancy age 41, Hugh age 21, John M. age 16, James H. age 11, Solomon J. age 9, William F. age 7 and Thomas age 5. Stephen and Nancy were born in North Carolina; all the children were born in Tennessee. This would indicate that Stephen came to Greene County sometime between 1845 and 1850. He was listed on the 1850 tax assessor's list in Greene County.

Stephen was also listed in the 1856 tax list for Greene County. He was assessed for 1 horse valued $40, 7 cattle valued $110 and land Section 25, Township 30, Range 22, no value given for the land. This is the land he obtained with patent #15235.

Stephen entered a preemption land patent #15235 for the SE ¼ of the SE ¼ of Section 25, Township 30, Range 22. Patent granted 15 December 1854. He also entered a land patent for the N ½ of Lot #1 of the NW ¼ of Section 31, Township 30, Range 21. Patent granted 1 June, 1859. This property is just south of the land in Section 25.

After Stephen's wife Nancy died, he married Juda (Judy) Elizabeth Scott. They had four boys: Andrew R., Henry G., Josiah and Willis E. born 1869. Willis is sometimes referred to as William in legal documents and newspaper notices, so caution is advised in researching his name.

On 26 December 1871, Stephen made his last will and testament and acknowledged his heirs. (43) The fractional part of S31, T30, R21 was bequeathed to his sons, Hugh, John M., James and Thomas. The land in S25, T30, R22 was bequeathed to his wife Juda Elizabeth and his sons, Andrew R., Henry G., Willis E., and Josiah.

On 5 August 1861, a number of local young men enlisted in the Union Army, including four of Stephen's sons, all who served in the 24th Missouri Volunteer Infantry. James Henry enlisted 5 August 1861 and was mustered out 14 October 1864. William enlisted 5 August 1861 and was mustered out 14 October 1864. Solomon J. enlisted on 5 August 1861. Solomon was mustered into service 14 October 1861 at St. Louis, Missouri. He was wounded at the Battle of Pleasant Hill, Louisiana on 9 April 1864 and mustered out of service 14 October 1864 at St. Louis. John Madison enlisted in Company A, 13 May 1862, but was never mustered into service. He was discharged 30 October 1862 for disability. (44)

William B. Dickens, Stephen's younger brother, came to Greene County later than Stephen. William B. Dickens was enumerated in the 1850 census of Bedford County, Tennessee. William was listed as age 36, laborer, born Tennessee, Nancy age 30, born Tennessee, Mary age 15, Lewis age 12, Thomas age 10, Madison age 8, Stephen age 5. All the children were born in Tennessee.

The family moved to Greene County, Missouri, before 1856 when William was assessed 1 horse valued $35, 4 cattle valued $70, 1 poll tax.

On 13 April 1856, Mary Dickens, William Dickens' daughter, married John Banfield (spelled Bunfield on record) in Greene County. (45) The couple was married by Joseph Headlee, J.P. (Justice of the Peace). John was the son of John and Tabitha Banfield. See Chapter 5, Banfield Family.

On 1 October 1860, patent #28.842 was issued to William B. Dickens for the SE ¼ of the SW ¼ of Section 25, Township 30, Range 22, consisting of 40 acres. Section 25 is immediately east of Section 26.

The 1860 census of Greene County, Missouri, Robberson Township listed William B. Dickens age 53, farmer, born Tennessee, real estate value $1000, personal property value $300, Nancy age 43, born South Carolina, Lewis age 20, James age 15, William age 13, Stephen age 10, Solomon age 8, John age 7. All the children were born in Tennessee.

On 30 December 1860, the oldest son, Louis Henry Dickens married Sarah E. Terrell. Sarah was the daughter of Thomas Terrell and Rachel Banfield. Rachel was the daughter of John and Tabitha Banfield. See Chapter 5, Banfield Family. The spellings Louis and Lewis appear in many documents, but Louis H. was carved on his tombstone, so it is believed this was the correct spelling.

On 5 January 1861, William Dickens made his will. Thomas Terrell and David McCurdy attested to the will in the Greene County Court, 6 January 1861, but the will was not found in the Greene County Archives. In a later document, Louis Dickens' probate file listed the heirs to William's property with dates when Louis bought out the other heirs. The heirs were Mary Banfield, Louis H. Dickens, James M. Dickens, William R. Dickens, Stephen P. Dickens, Solomon F. Dickens and John P. Dickens.

Louis H. Dickens, the oldest son, enlisted to serve in the Civil War. He enlisted into Company A, 24th Missouri Volunteer Infantry, on 29 September 1862 in Springfield, Missouri. He never mustered into service and was discharged 22 December 1862. He had earlier served in Company K of the Greene-Christian County Home Guards. (46)

Louis and Sarah Terrell Dickens had four children: Henry Allen born 1863, George W. born 1867, Oliver M. born 1870 and Walter born 1878. After Sarah's death, Louis married Judy Elizabeth Scott, who was the second wife of Stephen Dickens. This means that Louis married his aunt by marriage. Their children were: Clinton born 1882 and Linda born 1885.

On 26 December 1885, Judy E. Dickens and Lewis Dickens sold to Mary C. Bath, the SE ¼ of the SE ¼ of Section 25, Township 30, Range 22, except a small parcel 16 feet square described as a graveyard and the right of way to get into said graveyard. (47) The Dickens graveyard was identified by the Ozarks Genealogical Society as located 2 miles north of Springfield, about 100 feet to the East of Farm Road 151, across from Lakeview Lighthouse Church. The researchers indicated that there were two burials there, but no stones. Since William Dickens originally owned this property, it is probable that he and his first wife were buried there.

When Louis died in 1902, his real estate inventory included 80 acres, the S ½ of the SW ¼ Section 25, Township 30, Range 22. (48) This is the land which William B. Dickens had obtained by patents and which Louis bought from the other heirs. Louis left an inventory of dates when he purchased the land from other family members with a statement that the deed purchasing Mary Banfield Dickens' interest was burned in a fire before it was recorded.

The information is further confirmed by a lawsuit settled in Greene County Circuit Court on 3 October 1904. (49) Statements were entered which said that in 1868 Louis bought the share of his sister Mary Banfield and her husband John for $120. The house of Louis Dickens was destroyed by fire and the deed was burned. Mary and John were long deceased and many of their heirs were unknown and the heirs had no claim to the estate of Louis Dickens. (This lawsuit details family history and lists descendants of William Dickens).

An additional real estate asset was listed in the probate inventory. There were 20 acres in the N ½ of the NW ¼ of the NW ¼ Section 36, Township 30, Range 22, which Louis bought from Henry R. Chandler and Eliza E. Chandler 13 February 1882.

At the estate sale of Louis H. Dickens, most items of personal property were purchased by the Dickens heirs; however, A.J. Murray bought one red heifer for $14 and Will Murray bought a black calf for $6.75 and a red calf for $9.50.

Distribution of the estate was to Louis' children and grandchildren with the exception of heirs of Henry A. Dickens, who was deceased. It was stated that Henry's children were not entitled to a distribution because in 1897 Louis had advanced $400 to his son, Henry, and Henry accepted this in full as his share of the estate of his father.

Louis H. Dickens was buried in the Murray family cemetery. See Chapter Seven, Burials in the Murray Cemetery. It is curious that he was not buried in the Dickens graveyard, the area that was reserved in the 1885 sale.

The Civil War and Greene County Settlers

In an interesting sidelight to the above settlers, it seems that in 1861 several of the young men in the neighborhood were disposed to causing some commotion. The Grand Jury in Greene County issued three indictments at their January 1861 term. (50) No disposition of the cases was found in Greene County records, possibly because of the Civil War conflict.

A cause for disturbing the peace and damaging the property of Thomas McCurdy was brought against Samuel Foster, John Foster, Calvin Rector, John Dickens, James Dickens, William Dickens, Solomon Dickens, Reese Gott, Joseph Gott, Lee Roberts, James White, Lewis Banfield, John Daniels and Charles Thornhill.

Another indictment was handed down against James and John Dickens for disturbing the peace of Union Campground. The indictment alleged that the Dickens boys "willfully and maliciously and contemptuously disturbed and disquieted a congregation and assembly of people for religious worship by making noise, by rude behavior, by indecent behavior and by profane discourse."

The third indictment, for disturbing the peace of William B. Berry, was against some of the above named young men. It was interesting to see the family and neighbor relationships between the bondsmen and the perpetrators: Stephen Dickens, bondsman for William and Thomas Dickens, $100 bond for each; Samuel Foster, bondsman for John Foster $100; R. Q. Banfield bondsman for Lewis Banfield $100; William Jones, bondsman for Stephen Chandler $100; Epperson Thornhill and George B. Rabe, bondsmen for Charles Thornhill $100. (50) Some of these men enlisted to fight with the Missouri State Guards or the 24th Volunteer Infantry in 1861 and 1862. Perhaps that ended their enthusiasm for disturbing the peace.

The Battle of Wilson's Creek, just south of Springfield, was fought 10 August 1861 and the 24th Volunteer Infantry was in that battle. However most of the new enlistees named above had not yet been trained and they served later in the Battle of Pea Ridge. The history of the *24th Missouri Volunteer Infantry* stated that Charles Thornhill, son of Epps and Mary Thornhill enlisted with Company A, but deserted August 1861 in Rolla, Missouri. (51) He might have decided that the military was not for him, or he might have joined some of the young men in the community who went south and fought with the Confederate Army.

Following the Battle of Wilson's Creek, the Confederates were in control of the area and remained in Greene County until February 1862. According to Holcombe, "The first day of January 1862 saw Greene County under complete Confederate domination. Everything was done under complete marshal law. Property was seized for use of the army wherever it could be found." (52) General Price was forced to retreat from Greene County in February 1862 and many families followed him south, including D.D. Berry, the merchant mentioned in Chapter One.

The January 1863 term of the circuit court returned and cases resulting from the occupation were handled. According to Holcombe: "During the Confederate occupancy of Greene County many of the Union citizens had their property taken by Confederates, some of whom were also citizens of this county and owned property. After Federal authority was restored, suits were instituted against them by some of the Unionists living here whom they had arrested or whose property they had taken for military purposes. As personal service could not be had, notice of the suits was given by publication in the newspapers, which of course the defendants never saw, until long after judgment had been rendered by default and execution issued and served and their property levied on and sold." (53) This could have been the reason for the lawsuits against Samuel Lastley, page 50 and John Daniel, page 58.

References

(1) Colton, J.H. *The Western Tourist & Emigrants Guide*, (New York, 1844), 81, https://archive.org.

(2) Frederick B. Goddard, *Where to Emigrate and Why*, (Philadelphia, 1869), 290, https://archive.org.

(3) *An Index of the Springfield Land Office Sales Book 1833-1892*, Greene County Archives Bulletin #36, (Springfield, MO: Greene County Missouri Archives).

(4) Marsha Hoffman Rising. *Opening the Ozarks: First Families in Southwest Missouri 1835-1839*, (Derry, New Hampshire: American Society of Genealogists, 2005), 1203, ISBN # 1-59975-350-2.

(5) Return I. Holcombe. *History of Greene County Missouri*. (St. Louis: 1883), 143.

(6) Ingle, "Greene County Mills," a short publication found in Greene County Archives.

(7) Government Land Office Survey and Field Notes, at Greene County Archives.

(8) Greene County Missouri Marriage Book A.

(9) Cedar County Missouri Probate File, Samuel G. Lasley, September 1865.

(10) Greene County Missouri Deed Book I-301.

(11) Rising, *Opening the Ozarks* [note 4], 88.

(12) *Greene County Missouri Cemeteries, Volume X*, Ozarks Genealogical Society.

(13) *Missouri Still Births and Miscellaneous Records 1805-2002*, www.ancestry.com.

(14) Greene County Missouri Deed Book H-604.

(15) Greene County Missouri Deed Book I-292.

(16) Greene County Missouri Deed Book I-293.

(17) Greene County Missouri Deed Book S-378.

(18) Greene County Marriage Book A.

(19) Greene County Marriage Book B.

(20) Greene County Missouri Deed Book G-341.

(21) Greene County Missouri Deed Book J-522.

(22) Greene County Missouri Deed Book F-178.

(23) Greene County Missouri Marriage Book A.

(24) Greene County Missouri Circuit Court Case Book G-278.

(25) Greene County Missouri Circuit Court Case Book G-306.

(26) *History of Benton, Washington, Carroll, Madison, Crawford, Franklin and Sebastian Counties, Arkansas*, (Chicago, 1889), 1039-1040.

(27) J. Randall Houp, *The 24th Missouri Volunteer Infantry: "Lyon Legion,"* (Alma, Arkansas: J. Randall Houp, 1997), 244.

(28) "Missouri Marriage Records 1805-2002," www.ancestry.com.

(29) "Genealogies of Some Early Springfield Families," *Ozar'Kin*, Spring 1979, Ozarks Genealogical Society.

(30) "Missouri Marriage Records 1805-2002," www.ancestry.com.

(31) Greene County Missouri Deed Book F-440.

(32) Greene County Missouri Deed Book H-264.

(33) Greene County Missouri Deed Book H-687.

(34) Greene County Missouri Deed Book 29-31.

(35) Greene County Missouri Deed Book 33-106.

(36) Greene County Missouri Deed Book 30-146.

(37) Greene County Missouri Deed Book 34-285.

(38) Greene County Missouri Deed Book 199-357.

(39) Greene County Missouri Deed Book 119.

(40) Rising, *Opening the Ozarks [note 4]*, 77.

(41) Greene County Missouri Deed Book 46-582.

(42) *History of the Greene County Alms House, Book 1*, Greene County Missouri Archives.

(43) Greene County Missouri Probate File #2392. Greene County Missouri Archives.

(44) Houp, *24th Missouri Volunteer Infantry* [note 27], 247–250.

(45) Greene County Missouri Marriage Book B.

(46) Houp, *24th Missouri Volunteer Infantry* [note 27], 248.

(47) Greene County Missouri Warranty Deed Book 58-2.

(48) Greene County Missouri probate file #2488.

(49) Greene County Circuit Court Book 72, 358–360.

(50) Greene County Missouri Circuit Court Records, January 1861.

(51) Houp, *24th Missouri Volunteer Infantry* [note 27], 286

(52) Holcombe, *History of Greene County* [note 5], 401.

(53) Holcombe, *History of Greene County* [note 5], 456–457.

Chapter 5

The Banfield Family

The Banfield family is included because they had close connections with settlers listed in Sections 35 and 26 and because of marriages to members of the Murray family.

One of the early settlers in Section 36 was John Banfield. He and his wife came from Wilson County, Tennessee sometime around 1835. He purchased the W ½ of SE ¼, Section 36, Township 30, Range 22 on 3 January 1839. The patent #1340 was not issued until 20 April 1843. John Banfield served as a juror in the Randolph Britt murder trial in March 1839, but his family was not enumerated in the 1840 census.

On 11 July 1838, John Banfield made his will, which stated: "I, John Bandfield doe leave all his property in the hands of his wife Tabithy Bandfield than after paying his just dets and to devide with her children as she sees cose in give in her will." The will was witnessed by Louisa Carter, Polley Kidd and Joseph Hall and proven in 1840. (1) It seems strange that he would write his will in 1838 and then serve on the jury in 1839, but perhaps he had a chronic health problem.

On 11 July 1844, Tabitha executed her will asking to be buried next to her husband. She left 80 acres of land in the care of her son Qualls to be used by him for the care of her children until the youngest, Bradley became of age, at which time the property was to be divided equally between her sons, James, John, George and Bradley. In the event George and Bradley became 18 years of age, they were each to receive a horse, saddle and bridle. And in the event that Tabitha and Elizabeth Banfield, her two youngest daughters, were to marry, they were to be set up in housekeeping as her two daughters Rachel and Anne had been set up. Remainder of the property was to be retained by Qualls for his use and benefit. Qualls was appointed executor. The will was witnessed 11 July 1844 by William Warren and John H. Deeds. (2) There is a family relationship, not yet researched, between William Warren, Louisa Carter and Lucy Ann Warren who married Qualls Banfield.

The name Qualls, Qualles or Quarles Banfield was used in several documents. Marsha Rising noted a connection between the Banfields and someone named Quarles in Wilson County, Tennessee. (3) An entry on a public member tree on Ancestry.com noted that Qualls might be Tabitha's maiden name. No proof was given. He will be referred to as Qualls in this book, although later in life he was referred to as Roger.

John and Tabitha Banfield had the following children, all born in Tennessee.

1. Rachel, born 1817, married Thomas Terrell on 31 January 1841, by Sol Owen. (4) Rachel and Thomas lived in the neighborhood and are mentioned in Chapters 5 and 6.

2. Anne, no birthdate known, was married to Lemuel Williams, 3 September 1840 by Sol H. Owen CCJ (County Court Justice). (5) An earlier marriage is suspected for Lemuel Williams as he was listed in the 1840 Greene County census with children, but no female of the age to be his wife. He was found in the 1852 Missouri State Census for Greene County living in Township 30, Range 29. On that census list there is one male under 10, 2 males 10-18 years and 1 male 45 years and older; 2 females under 10, 1 female 10-18 years and 1 female 45 years and older. There are references to Lemuel Williams in early business records of Greene County, but he is not found in the 1860 census.

3. Qualls, born 14 March 1820. He married (1) Lucy Ann Warren 29 August 1843 (2) Huldah Murray 30 November 1873 (3) Aulsie McGraw 5 November 1879.

4. James, born 1825, married Elizabeth Carter, 20 May 1844 in Greene County, Missouri. (6) She was the daughter of James and Sarah Carter, relationship established in her father's probate records. (7) There is no family relationship between this Elizabeth Carter and Hazen Carter who married Elizabeth, daughter of Qualls Banfield. Hazen was born in Vermont in 1841, the son of Jeremiah Carter.

5. Tabitha, born 1826. She married James Hagen on 12 January 1847 in Greene County. No further record of her is found in Greene County.

There is a public member tree on Ancestry.com which states that she was living in Lotts Creek, Iowa in 1856 and died in 1906 at Lotts Creek. There is a picture of a stone in Oak Dale Cemetery, Lotts Creek, Iowa for James Hagans, born 1809, died 7 September 1863. Mentions a wife Tabitha. This could be Tabitha Banfield.

6. John Banfield, born 1827, married Mary E. Dickens on 13 April 1856 in Greene County, Missouri. She was the only daughter of William B. Dickens.

7. George: No record in Greene County.

8. Bradley: Possibly listed in 1850 census, Robberson Township.

9. Elizabeth: No record in Greene County.

In the 1850 census of Robberson Township, Greene County, Missouri, dwelling #602 is that of Quarles Banfield, age 37, born Tennessee, value of real estate $300, Lucy A. age 22, born North Carolina, Lewis H, son age 7, E.F. daughter, age 5 and Mary J. daughter, age 1. The head of family #603, is 24-year-old Amanda Lanham, who appears to have no relationship to the Banfield family. Living with her is William B. Banfield, age 14, born in Missouri. It is possible that he is the Bradley listed in Tabitha's will. No other records pertaining to Bradley have been found in Greene County.

Qualls name is listed often in Greene County records, serving on grand juries and as a witness in civil cases. In March 1845, he was called as a witness for the state in the case of an altercation between Charles Baker Owen and defendant Bennett Deeds. Qualls' testimony: "On yesterday, went to a meeting at William Fulbright's on the Sac. Charles Baker Owen, the defendant and others were there. We was a talking about running horses." Owen and Deeds exchanged words. Owen thought Deeds called him a "goddamned rascal." Owen tried to find a rock, Deeds pulled a knife. Deeds cut Owen two or three inches deep on the left shoulder, near the collarbone, but it was not life threatening. (8) There was no disposition of the case found in Greene County records. The case may have been filed because Charles Baker Owen was the son of Judge Solomon Owen.

Qualls is first found serving as a juror, 25 August 1854, in the criminal trial of Willis Washam, which resulted in a guilty verdict and the hanging of Washam. The jury was composed of some early northern Greene County settlers: Ezekiel Cook, Qualls Banfield, William White, and James S. McQuirter.

In the Circuit Court of the March session, 1858, Samuel Burlin Cate was accused of selling intoxicating liquor to R.L. Banfield, for twenty-five cents, violating the dram law. No disposition of the case was found.

Then in the March session of Circuit Court, 1859, R.Q. Banfield was accused of contempt of court for failing to appear as the state witness in the case of John Daniel, defendant, on charges of gambling. This case was described in the section on Tapley Daniel, Chapter Four. It is possible that Qualls did not want to appear as a witness against his neighbor's son. Once again there was no disposition of the case.

The 1860 census of Robberson Township, Greene County, Missouri, listed F.T. Banfield (R.Q.) as 35 years, farmer, born Tennessee, real estate value $2000, personal property $2000, Lucy age 30, born North Carolina, Westley age 14, Elizabeth age 13, Nancy age 11, Columbus age 8, Quarles age 6, John age 2. All the children were born in Missouri.

The family living next to R.Q. is John Banfield age 33, born Tennessee, farmer, mill & c. gin, value real estate $1000, his wife Mary age 23, born Tennessee and Louise age 2, born in Missouri. (John is the brother of Qualls; Mary is the daughter of William B. Dickens.)

In November 1863, the United States Civil War Registration Records 1863-1865, show a Valls Banfield (Qualls) born 1820, Tennessee, age 43, farmer. He was registered by Captain Bodenhamer, Provost Marshall, although there is no evidence that Qualls served in the military.

Qualls' brother, John, served in Company A, 24th Missouri Infantry. He enlisted 5 August 1861 in Springfield, Missouri. The record stated he was 37 years of age, occupation miller, born Wilson County, Tennessee, enlisted for a period of three years. His eyes were grey, hair dark, complexion dark, height was five foot ten inches. He was assigned to the Provost Office in May 1862 in Springfield, Missouri, and in July and August was assigned as assistant detective police at Springfield. He was mustered out 14 October 1864 and paid $100. (9)

11 November 1864, John and Mary Banfield sold twenty acres of the SW ¼ of the NE ¼ of Section 35, Township 30, Range 22 for $500 to William Jones. (10) William Jones would later be buried in the Murray cemetery. This is the same property John Banfield bought from James McQuerter in 1857. It was about this time that John and Mary left Greene County.

In the 1870 census of Owen Township, Warrick County, Indiana, there was a listing for John Banfield age 46, farmer, born Tennessee living with Mary E. Banfield age 32, born Tennessee and children Louisa age 13, Nancy J. age 9, John age 7, Rachel age 5 and James A. age 2. Rachel and James were born in Indiana; the other children were born in Missouri. The dates of birth for Rachel and James would correspond with the time when John was mustered out of the Army and the family left Missouri.

There is further confirmation of the above information, which came from a case in Greene County Circuit Court 21 September 1904. (11) In the case Elizabeth Dickens, Linda and Clinton Dickens were plaintiffs against descendants of Mary and John Banfield, deceased. It states, "In 1868, Louis H. Dickens bought the share of his sister Mary Banfield and her husband John Banfield [they were shares in the William B. Dickens estate] ---the dwelling house of Louis Dickens was burned and the deed from Mary Banfield and her husband was destroyed." The case was settled in favor of the plaintiffs. See Louis Dickens history, Chapter 4.

The 1870 census of Robberson Township showed Roger Q. Banfield age 48, Lucy A. age 33, Columbus 18, John age 11 and Lucy A age 9, living in household #95. The family next door was Ann Warren age 69, who must have been Lucy's mother, born North Carolina with Melissa age 21, born North Carolina and James age 19, born in Tennessee. Also in that family was Fanny age 51, black, born North Carolina, personal property valued at $300. Fanny was not enumerated in the 1850 or 1860 Greene County slave schedules, but it can be inferred that she was a slave in the Warren family and came to Missouri with them. She is buried in the Banfield Cemetery with Qualls and Lucy. The cemetery and farm are now a part of

Lost Hill Park, owned by the Springfield-Greene County Park Board. Fanny must have been highly regarded by the family. Her stone reads: "Fannie, born 1816, Halifax County, Virginia. Died February 16, 1888. Praise waiteth for the Fannie. O God of Zion and unto thee shall the vow be performed."

Lewis, the oldest son of Qualls and Lucy, married Nancy Giboney 19 July 1865. Elizabeth married Hazen B. Carter 17 September 1863. Mary married Samuel Dishman, 3 August 1865 and Lucy married Benjamin Potter. Information on Lewis and Nancy Banfield and Elizabeth and Hazen Carter can be found in Fairbanks and Tuck, Volume I. (12)

Lucy Ann Warren Banfield died 4 July 1873 and was buried in the Banfield Cemetery. The only stones with inscriptions are those of Lucy, Qualls and Fannie.

After Lucy's death, Qualls married a neighbor, Huldah Murray, widow of David Murray. It was quite surprising to find an ante-nuptial agreement signed 29 November 1873. (13) They were married 30 November 1873. Because Huldah had five children, she apparently felt it was necessary to protect her children's inheritance.

The terms of the agreement were: (1) Huldah agreed to place all her personal property in the house to be occupied by both parties; she also agreed to place her livestock and farming utensils with the party of the second part (2) the title to one undivided half of all personal property, livestock and farming implements shall remain Huldah's property and shall be used for her separate benefit and that property shall in no way be used to pay for the debts of Banfield (3) in case of death or dissolution of the marriage, she shall be entitled to all of her household and kitchen furniture and one half of all properties acquired during the marriage (4) in case of death, both parties agree that the properties of the parties shall go to the heirs of each party (5) that Huldah shall have the right to go to court in her own name or that name of her best friend.

Qualls and Huldah had a son, Roger Lawson, born sometime between 1875 and 1877. The census taker listed him as being five years old in the 1880 census, but Roger later reported different birth dates. Qualls and Huldah were divorced May 1877, divorce award to Qualls. (14) They re-married 8 February 1879. (15)

Huldah Murray Banfield died 5 July 1879 and was buried in the Murray Cemetery, next to her first husband, David Murray, with the name Huldah Murray on the tombstone. See Burials in the Murray Cemetery, Chapter Seven. Her husband, Qualls Banfield, declined to serve as executor of her estate and T.S. Wilson, a neighbor, was appointed executor. (16)

The number of household items divided between her heirs indicated that her house was fully furnished. Were these items purchased after she built the brick house she wanted in Missouri, completed in 1878, or had she acquired them earlier? If so, where was a house large enough to accommodate them? There were no family stories to indicate whether Qualls lived in her home on the Murray farm or she lived in his home. The ante-nuptial agreement indicated that she was to place her belongings in his home, but that would have been terminated after the divorce. Research has revealed no answers to these questions.

Huldah's children Zelotus, Jasper, William, Andrew and Evalina each received a bed and bedstead valued at $25, little Roger received a trundle bed. A bureau, table, chairs, rocking chairs, dining room set, stove, tableware, spinning wheel and sewing machine were divided among the children. The boys were given mules, mares, cows and hogs. Since none of the heirs were married, David's children remained on the Murray home place and Roger remained with his father.

In the 1880 census of Robberson Township, Jasper Murray age 22 was listed as head of the household with Zelotus age 19, Andrew age 14, William age 12, Evaline age 9. There was an additional family living there, headed by Fred Hofman, who was listed as farm laborer. Apparently he was helping with the farm and his wife was keeping house for the family.

The boys must have been concerned about their little sister's education and proper clothing. Two receipts were found. The first: Received of T.S. Wilson, Curator of Evalina Murray, this 28th day of August 1880, eighty cents for cutting and making two dresses for the said Evalina Mury. Signed by Mrs. Hoffmann. This was the same Mrs. Hoffman who was living with the family in the 1880 census.

The second receipt was: T.S. Wilson, Curator of Evalina Murry to Mifs. D. J. Kirkpatrick To 3 month's tuition commencing in March and ending in June

1880. $1.50. Received payment in full of the above account this 26th day of June 1880. Signed, Dorcus J. Kirkpatrick. Dorcus was not found in the 1880 census, but was enumerated in the 1870 census living in the home of Martha Gurley. A woman named Martha Kilpatrick, age 36 with daughters Ann age 13 and Dorcus J. age 9 were living with Mrs. Gurley. Neighbors were Hosea Mullings and Rachel (Banfield) Terrell.

In the 1880 census of Robberson Township, Roger Banfield age 55, was living with his wife, Alice age 40, James Wilson age 17, step-son, Roger age 5, son and Fanny Banfield age 60, black. On 5 November 1879, Qualls had married Mrs. Aulsie (Alice) McGrew. (17)

On 6 May 1887, Roger Q. Banfield wrote his will. (18) It is interesting that he used the name Roger more often in his later years. He asked that the 200 acres he owned in Robberson Township be sold and the proceeds divided equally between his wife, Aulsie and children: Elizabeth Carter, Mary Dishman, John Banfield, Lucy Ann Potter and Roger Lawson Banfield. His two sons, Lewis and Columbus were given one dollar each because he had heretofore given them 40 acres each.

Gravestones of Roger Q. Banfield, and of Lucy A. Banfield

He asked that his personal property be sold, with the first $200 given to Roger Lawson Banfield as part of the proceeds of his mother's estate. The remainder from the sale was to be divided between the other heirs. He gave a house and two lots in Bethesda, Greene County, Missouri to his wife, Aulsie. The proceeds of a sale of a lot in Kellets Addition, city of Springfield, were to be divided equally among the heirs.

He named his wife as executrix of his estate and asked that when the homestead was sold, one acre be reserved "where my father and mother and wife are buried." This is the Banfield Cemetery, now a part of Lost Hill Park.

Although T.S. Wilson was initially appointed guardian of Roger Lawson Banfield, on 28 September 1889, Andrew J. Murray, Roger's half-brother was appointed guardian. In the October 1890 term of the Probate Court, A.J. asked to be discharged from that duty. J.S. Owen became curator of the estate until 21 November 1896 when he stated that Roger L. Banfield is "now above the age of 21 years." This would indicate that Roger L. was born in 1875 and would correspond with his age on the 1880 census.

Roger Lawson was not living with either a Murray or Banfield family in the 1900 census. There was an "I.W." or "L.W." Bandfield living with the Charley Walters family in Polk County, Missouri. This male Bandfield was 24 years old and was working as a day laborer. There is a possibility that this is Roger. His name has been listed in various ways in the census reports and the age would correspond to his date of birth in the 1880 census reports. There was a family connection to Polk County. Roger's half-brother Louis and wife Nancy were living in Polk County in 1900.

In the 1910 census of Colfax Township, Raton City, New Mexico, there was a Robert L. Banfield, age 33, born Missouri, living with his wife, Allie age 27, and children, Gladys age 7, Bernice age 2 and Urban age 1. All the children were born in New Mexico. The census stated that he was an engineer with the railroad and owed his own home, which was mortgaged. Although Robert answered that his parents were born in Missouri (the same answer as I.W. or L.W. in the 1900 census), I believe this is Roger Lawson Banfield.

Roger L. Banfield's World War I registration was made in Tucumcari, New Mexico where he stated that his date of birth was 26 December 1878. Registration was signed "Roger L. Banfield," dated 9/12/18. He said he was an engineer with the railroad, closest relative was Allie Banfield. Physical description was "short, medium build, blue eyes, brown hair."

In the 1920 census of Precinct One, Tucumcari, New Mexico, there was an R. L. Banfield listed as age 41, born Missouri, parents born Missouri, employed as a locomotive engineer. Living with him was his wife, Lobetta A. age 36, Burnice daughter age 12, Urban son age 10. The census on Ancestry.com listed his first name as A.l. or H.L.

In the 1930 census of the county of Quay, New Mexico, Lawson C. Banfield was listed as a roomer, age 53, divorced, age 24 when first married, working as a locomotive engineer. This time he did answer that his father was born in Tennessee and his mother was born in Ohio, which is correct.

A death certificate from New Mexico stated that Roger Lawson Banfield died 20 December 1943 of arterial sclerosis and hypertension. His occupation was listed as railroad engineer with the Southern Pacific Railroad. The informant was his wife, Maude, who gave his birthdate as 24 December 1877. She said he had resided in Quay County for 39 years. He was buried in the Tucumcari, New Mexico cemetery.

Roger Lawson Banfield is listed in this section because his father was Qualls Banfield. The Murray family had little contact with Roger Lawson. He occasionally visited Missouri and saw his half-brothers or their families, but Murray family members did not seem to know much about "Uncle Lawson." The history of the Murray family as it relates to the Greene County area will be covered in Chapter Six.

References
(1) Greene County Missouri Will Book A-710.
(2) Greene County Missouri Will Book C 45-46.
(3) Marsha Hoffman Rising. *Opening the Ozarks: First Families in Southwest Missouri 1835-1839*, (Derry, New Hampshire: American Society of Genealogists, 2005), 98. ISBN # 1-59975-350-2.

(4) Greene County Missouri Marriage Book A.

(5) Greene County Missouri Marriage Book A.

(6) *Missouri Marriage Records 1805–2002*. www.ancestry.com.

(7) Marsha Hoffman Rising, *Greene County Missouri Probate Records 1833-1871*, (Greene County Archives) Minute Book B, 122.

(8) Greene County Missouri Court Docket, 31 March 1845.

(9) J. Randall Houp, *The 24th Missouri Volunteer Infantry*, (Alma, Arkansas: J. Randall Houp, 1997), 233.

(10) Greene County Missouri Deed Book D-546.

(11) Greene County Missouri Circuit Court Record Book 81, 111–114.

(12) Jonathan Fairbanks and Clyde Edwin Tuck, *History of Greene County, Volume 1-2*, (Indianapolis, Indiana: A.W. Bowen & Co., 1915).

(13) Greene County Missouri Deed Book 29-429.

(14) Divorce Records Greene County Circuit Court Book P, 503.

(15) Greene County Missouri Marriage Book D, 382.

(16) Greene County Missouri Probate File, Huldah Murray, #555.

(17) Greene County Marriage Book E-10.

(18) Greene County Missouri Probate File, Qualls Banfield, #472.

Chapter 6

David Murray Comes to Missouri

Photo of David Murray, printed from a tintype, date unknown

David Murray came to Missouri almost thirty-five years later than the people listed as first settlers along Boonville Road. His interactions with settlers Tapley Daniel, Samuel Austin, the Dickens family and the Banfield family were discussed in Chapters Four and Five.

The *Wyandot Democratic Union* newspaper, Upper Sandusky, Ohio, on 22 August 1867, printed the following: "Personal. Our esteemed friend, D.C. Murray, Esq. started, with his family, on last Monday, for his new home in the West, five miles from Springfield, Mo. We, in common, with the citizens of this county, regret exceedingly to part with the Squire, for he was one of our most influential citizens.

We truly hope he may find his new home pleasant, profitable and agreeable to all his expectations." (1)

Why did David Murray move his family from Ohio to Missouri? One relative in Ohio said that he moved to a state with warmer climate because of his poor health. Another mentioned that he wanted a farm ideal for raising horses. He sold 200 acres in Wyandot County, Ohio to George Kear, a neighbor. That county had slightly rolling land with dark, loamy soil. Richland Township, where David's farm was located, was so named because of the "rich land." He moved to Greene County, Missouri, to the farm that he purchased from Tapley Daniel, located on a river bottom, hilly and covered with rocks.

A letter from David's brother, Z.G., who stayed in Ohio stated, "Hamilton (David's son-in-law) has just returned from your country and he is not impressed with it. He says you may wear out your wagon tires on the rocks." In another letter, Z.G. expressed concern for his brother and asks, "Are you making it all right? We have heard that you have to haul wood to make a living." One wonders if the new land lived up to the expectations. He apparently bought the land sight unseen.

Shortly after the Civil War, people became interested in moving west. They scouted newspapers for investment opportunities. In the *Springfield Missouri Weekly Patriot*, September, 1865, J.H. Creighton and Sons, real estate agents were advertising the sale of property in Greene County. J.H. Creighton advertised that he was "late of Ohio." Perhaps David learned about land for sale in Greene County from an advertisement in an Ohio newspaper. An article from the same newspaper, *Springfield Missouri Weekly Patriot*, on 12 October 1865 stated that immigration was increasing to the Midwest. The article stated, "Dispatches of the 3rd [of October] from Springfield, Illinois state that the time of emigration from the Middle States to Missouri seems to have set in in earnest. Scarcely a day passes in which some twenty to thirty families pass through this state in route to the latter state. Some thirty eight wagons loaded with emigrants passed today."

In the 12 April 1866 edition of the *Springfield Weekly Patriot* an article is reprinted from the *New York Tribune* extolling the virtues of Missouri. It stated, "The future of Missouri must be glorious. No other state presents so many and so varied attractions.... All this point men of small means as well as men of capital and enterprise to Missouri."

The *Springfield Weekly Patriot* in the 26 April 1866 issue describes the future of Springfield after the railroad is extended to the town. "When the Southwest Pacific Railroad is completed to our city, Boonville Street will be the main avenue… the great artery of business enterprise…and will make Boonville Street the Broadway of Springfield." These glorious descriptions of Missouri and Springfield would make any enterprising citizen want to settle in Greene County Missouri.

David's brother, Z.G. wrote that several of their neighbors in Wyandot County were traveling west to look for land in Iowa, Kansas, Nebraska and Indian Territory. Z.G. thought he had a buyer for his farm and asked David to find a farm that he could purchase in Greene County. Unfortunately the man who was planning to buy Z.G.'s farm could not sell his farm in Pennsylvania, so the deal fell through.

The pattern in David's family had been to move toward greater opportunity. His father, also David, had moved from New York to Ohio after 1816 when that state opened up for settlement and perhaps David, in turn, saw the availability of cheap land in Missouri as an opportunity for his four sons.

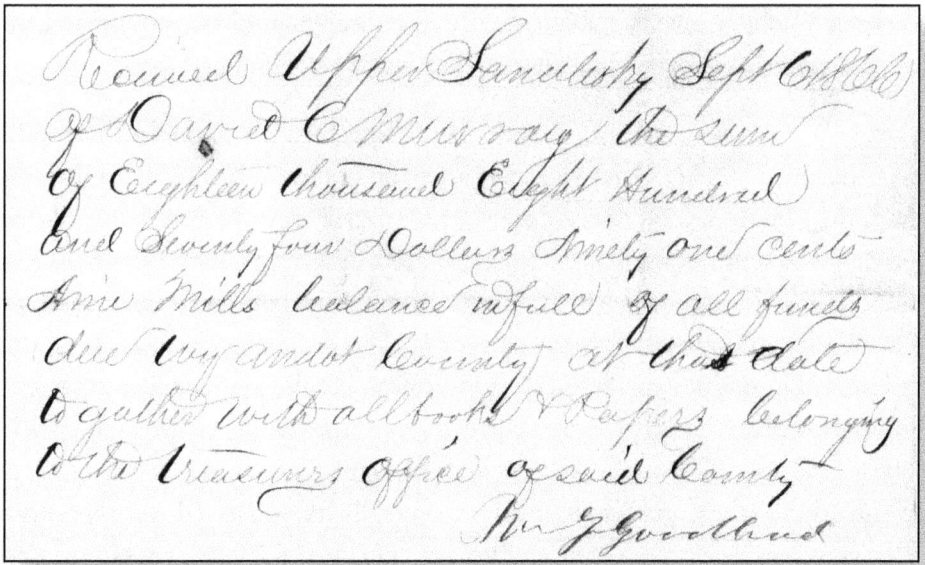

Receipt for papers and money due Office of Treasurer, September 1866

David apparently kept in contact with friends from Ohio and knew many people there due to his years as an elected official. He served as Justice of the Peace for Salem Township 1856–1858 and 1859–61. He was elected Treasurer of Wyandot County in October 1861 and re-elected in October 1863, serving to the end of the Civil War. His daughter Ezenith said that he dressed as a tramp and traveled by train to Columbus to carry pay for the soldiers.

On 6 September 1866, David Murray turned in all the papers and money due Wyandot County. From the dates on small receipts found in David Murray's papers, it appears that David may have loaned money to individuals who could not pay their taxes during the war. Ezenith, his daughter, said that when he left office he had less money than when he took office.

On 13 August 1867, David and Huldah sold 200 acres to George W. Kear for $7150, excepting one eighth of an acre used for burying purposes. This may have been where Susannah Long, David's first wife was buried. A resident of Wyandot County, who was contacted in 1985, said that there was a cemetery located on David Murray's property and that the owner had used tombstones for the floor of his barn. There is no way to verify this story. The Justice of the Peace examining Huldah for her dowry rights was her brother-in-law, Z.G. Murray.

A letter David received 12 December 1870 indicated an old friendship and the desire by the writer to purchase new lands in the West. His friend writes:

"I want to visit your place this winter if nothing happens to prevent me and I just thought I would bother you with a few questions. Is there still an opening in Springfield for a good or possible atty. Like myself, or is it full. What can good unimproved lands be bought for....I have been in Kansas City for the last eight months but have not moved my family yet. I have some boys and want to get some land for them near where I am to make my final house....I want eventually to turn my attention to the stock business and I want to get the best climate, water, etc. for that purpose." He writes of land he owns in Kansas and Iowa, which he would like to exchange for lands in Southwestern Missouri. Then he asks, "What kind of winters do you have and how long do you generally have to feed. Is your country a blue grass region naturally. How are times financially this fall. How is business, on the wane, or brisk. I think we are on the eve of a money crunch and I look for very hard times akin to those of '57." Signed A.M. Jackson. (2)

One wonders how David made contact with Tapley Daniel who sold him the 320 acres for $6000, on 5 September 1867. The newspapers were advertising good farmland for $10-15 per acre, yet David paid $18.75 per acre for hilly, rocky land.

And what did David and Huldah think when they arrived at their new home? They had left a large Italianate brick house in Ohio, but there was no brick house in Missouri.

Photo of the David Murray home in Ohio taken in 1985

Tapley Daniel and his wife lived in a house erected on the Boonville Road as indicated by a symbol on the early maps for a house located between the two Sac Rivers, but it would have been a frame cabin, similar to those described by Mr. Richardson in his ride across Missouri on the Butterfield Stage. Was Huldah disappointed to see the Daniel house, recalling the home she had just left in Ohio?

There are several letters in the David Murray papers that indicate a strong friendship between Murray and Daniel. Tapley signed his letters, "Your friend" or

"Your true friend." He asks for David's help in mediating a dispute with Qualls Banfield. Following is the complete account of the dispute first mentioned in the Tapley Daniel story, Chapter Four.

In July 1865, Roger Q. Banfield sued John Daniel (Tapley's son) for wrongfully taking his horse valued at $135 and clothing worth $40. (3) John Daniel was not a resident of Greene County. Later in July 1865, the court ordered the sale of two mares and a filly in Tapley Daniel's possession, with proceeds to go to Banfield. (4) In the Circuit Court Case Book, August 1866, the court decided that the plaintiff Banfield was the injured party and ordered the money received by the sheriff from the sale of property go to the plaintiff.

On 27 February 1866 Tapley Daniel filed suit against Quarles Banfield and John A. Patterson, sheriff, for wrongfully taking his horses and asks for damages of $700. Tapley's attorney was John S. Phelps, who had returned to the practice of law following the Civil War. The defendants responded that they took the horses by virtue of a writ of attachment from the court and that at no time did the plaintiff Daniel claim the horses as his property or attempt to forbid the sale.

The case dragged on. 31 July 1869 in Florence, Morgan County, Justice of the Peace David R. Davis took testimony from John Daniel who stated "that he did not own the property no way. He took his horses with him when he left the state on the 12 day of February 1862 and he did not return to the state until 1868 and that during the time he was absent from the state he had no horses in Greene County." (5)

There was testimony from neighbors in Greene County who said that Tapley had told them he was keeping the horses for John while he was gone, but Tapley, John and Tapley's daughter, Lucinda all testified that the horses belonged to Tapley. The case was finally settled out of court with costs charged to the defendants. (6)

There is no mention in the court records about David Murray's involvement, but from Tapley's letters, David must have talked with Qualls and worked out a solution.

In a letter from Florence, Missouri 8 January, 186- [the right edge of the letter is missing] Tapley asks David Murray "to go and tell Col. John S. Phelps that if the weather remains cold, I cannot be there and he will have to put off—[part of the letter missing] or do the best he can with it." This was puzzling until the Greene County Court case #3834 was found, as well as Tapley's other letters in David Murray's private papers.

In a letter December 1870, Tapley wrote, "I want you and Hardy White to make a compromise of that law suit if it takes all the judgment that I get ..." The completion of the sentence is illegible, but the letter is signed, "Your true friend until death."

A letter dated 16 April 1871 said, "As regards settling with Banfield, if he will pay you one hundred and fifty dollars and pay all costs and lawyers fees that I will settle with him." As noted earlier, the case was settled out of court with the defendant (Banfield) paying court costs.

24 May 1871, Tapley wrote again, "I was very well satisfied with the way you made the compromise with Banfield, pay yourself and send me the rest." From records found in Ohio, David Murray was very comfortable dealing with court cases and helping friends in trouble. His name was frequently found as executor in probate files and as a guardian.

Not much is known about David and Huldah's new life in Missouri except for letters written to David by his brother Z.G. in Ohio. Z.G. writes about old friends and neighbors in Wyandot County and discusses county politics because both brothers were active in the Democratic Party. Z.G. sends money collected on loans that David had made and money collected annually from George Kear on the sale of the farm. On 18 March 1871, Josiah Gibson wrote from Ohio regarding money and interest he owed David and thanked David for helping "the boys get along with their spring work." He notes another Ohio neighbor starting for Missouri in a few weeks planning to re-locate in Missouri.

I have never heard any family stories about David's religious affiliations, but he must have been an officer in a denomination not mentioned in a letter from a J.M. McDonald, Lebanon, Missouri, dated 29 November 1871. McDonald wrote that he sent minutes to David last week and apologized for the delay due to problems with the printer. It is obvious they are minutes relating to a church meeting because he noted, "They put elders in place of elder, Wilson in place of Wisdom on the committee of finance and low in place of law on a circular." He then stated that, "God has blessed us with three accessions to the church, but has deprived us of our dear brother Elisha Turner." David's Bible contained no family genealogy, only three invitations to funerals in Wyandot County, names that have not been found in any other papers.

David Murray was living in Robberson Township on the 1870 Greene County census. He was age 48, a farmer, with real estate valued at $6000 and personal property $1200. His wife was Huldah M. age 31, daughter Ezenith age 18, Jasper age 12, Zelotus age 9, Andy J. age 5, William age 3 and Evalina age six months. All were born in Ohio except Evalina, who was born in Missouri. It stated that she was born November 1869, a date at odds with her death certificate and her tombstone. Ezenith was David's daughter from his first marriage to Susanna Long, who died sometime after 1852. Huldah and David were married 13 December 1855 in Salem, Wyandot County, Ohio. He was 32; she was 16.

Picture of Huldah Murray, date not known

On 7 April 1872, David C. Murray made his last will and testament. He gave all his property to his wife Huldah during her natural life and after her death the remainder to the children of his wife Huldah: Jasper, Zelotus G., Andy J., William and Evalena.

Second, he gave $100 each to the children of his first wife: Jerema Sterling, Azenath A. Sell and Isaiah Murray, the sums to be paid from a promissory note paid by G.W. Kear of Wyandot County Ohio for $700, due in 1875.

His wife Huldah was appointed as sole executrix of the estate. He authorized her to sell the S ½ of the NW ¼ of the SE ¼ of Section 26, Township 30, Range 22 in Greene County, Missouri. This was the property that David and Samuel Austin purchased from the railroad. See Chapter 4. He also directed that his minor children were to have a home on the homestead farm with his wife during their minority.

David Murray died 9 April 1872. Inventory of his estate was made by George Faul and H.C. White, probably the same Hardy White referred to in Tapley Daniel's letters. (7) It is surprising that his personal property was valued at only $951.50, which included 12 horses, 5 cows, 31 sheep, 30 hogs and some household items. His assets were 320 acres which had been purchased from Tapley Daniel, land in the NW ¼ of SE ¼ of Section 26, Township 30, Range 22 purchased from the railroad, balance on a note from G.W. Kear of Wyandot County, Ohio, a note payable from S.H. Owen for $450 and a note from John and Samuel Austin for $180. The note was made 11 November 1871, between Samuel Austin and David Murray for the property purchased from the Pacific Railroad Company. The quarter was to be divided equally between the parties, giving each 20 acres. The property was purchased in the name of David Murray who would obtain the deed when full payment was made 3 August 1873. Sale of this land by Huldah M. Banfield in 1874 was noted in the section on Samuel Austin.

There was no mention about notes payable from individuals in Wyandot County and it is doubtful that the money which David had loaned was ever repaid.

On 30 November 1873, Huldah Murray married a neighbor, Qualls Banfield. There is no family history to indicate if she lived in Banfield's house or if he lived in her house, but by the terms of the ante-nuptial she was to place her personal household property in the home of Banfield. The 1879 inventory of her estate indicated a considerable increase in personal property over the property listed in David Murray's estate. It also indicated that the personal property was still located on the Murray farm. Details of Huldah's estate inventory are found in Chapter Five, The Banfield Family.

Sometime between her divorce and re-marriage to Qualls, Huldah began to build the house she wanted on the Murray home place, modeled after the home she left in Ohio. In October 1877 she purchased wood and paint from S.W. McLaughlin, proprietor of Springfield Planing Mill and continued purchasing shingles, windows, flooring, doors and lath. Her final bill from McLaughlin was 1 November 1878 and was billed to Mrs. Huldah Murray. Total charged by McLaughlin: $377.

There were also individual receipts, undated, for wagonloads of brick, which, according to family stories, were hauled by her sons, one wagonload at a time. These little receipts totaled 26,301 bricks; however there are two lines on receipts that are blurred, so she may have purchased additional bricks. It is strange that she finished the house in November 1878 and then re-married Banfield in February 1879.

The house is an Italianate style popular during the mid-1800s. There were house plans published in the *Canada Farmer Magazine* that look much like Huldah's house.

Plans were available by mail order and the author reminded readers: "It is the common practice of some of our farmers to take all their meals in the kitchen, this is a habit which marks a low state of society.... Our agricultural population should not scorn comfort and refinement. Every grace that belongs to rural life, should be found among the daughters of our farmers." (8) The chief features of the Italianate farm house are: Low pitched or flat roof, balanced symmetrical shape, two or three stories, overhanging eaves or cornices, a cupola, tall, narrow double paned windows and arches above windows or doors.

The following photograph was taken circa 1900 and shows the similarity of architecture with the house that appeared in the *Canada Farmer*. The exterior has been changed very little. Shown in front of the house are Z. G. Murray, son of David and Huldah, and his children: Luther, Walter, Dorsey and Mamie.

Early Settlers Along Boonville Road

Italianate farm house published in Canada Farmer Magazine, *1865*

Photo of Huldah Murray's house circa 1900

Sometime after the 1880 census when Jasper Murray was listed as head of the household and before November 1881, Jasper Murray was married to Maggie LNU, as evidenced by her appearance before a Greene County notary on 3 November 1881 to be examined as the wife of Jasper Murray. On 3 December 1881, Jasper Murray and Maggie Murray, his wife, executed a quitclaim deed for Jasper's part of the Murray farm. (9) He was paid $800 by Z.G. Murray; however the deed was not filed until 10 September 1883.

Receipt in divorce settlement between Maggie Gay and William Gay

The above receipt was found in the Murray papers and is a total puzzle; perhaps it could give a clue about Maggie. The receipt was a payment from William Gay to Maggie Gay in a divorce settlement, final decree dated 16 November 1881. She was awarded the divorce after the defendant "refused to obey any of the obligations imposed on him by his marital obligations to the plaintiff."

In the 1870 census, Joshua Gay, with a son, William, was living next to the family of Solomon Owen, a neighbor to the Murray family. William was still living with his father's family in the 1880 census, so his marriage to Maggie did not last long. No marriage record for William and Maggie has been found in either Greene or Polk Counties. The dates of the above payment and the execution of the quitclaim deed are so close as to raise the question whether this could be the Maggie who married Jasper Murray. And, if there was no relationship, why would the above receipt be in the Murray family papers? Unfortunately no marriage record for Jasper Murray and Maggie LNU has been found in either Greene or Polk counties.

A family member said that Jasper "sold out" and left the country. He returned a few years later minus his wife. When asked about his wife, he reportedly responded, "I don't know that's any of your business." Jasper must have returned to Greene County minus Maggie in 1885 because on 16 October 1885 he appeared before a notary and acknowledged that he executed the quitclaim deed in 1881. (10)

Jasper was listed in the 1900 Greene County Missouri census, age 40, widowed, working as farm hand, living with his brother Andy J. Murray. In the 1910 census, he was living with Robert and Nancy James, age 50, working as hired hand, divorced. In the 1930 Greene County census, he was living with his niece, Ollie Owen and her husband Jerome. He was age 60, widowed. There were family stories that Uncle Jap lived with one family member and then another, working for a time and moving on. When he was not working, he was reading his Bible. Jasper died 23 January 1940 and is buried in Greenlawn Cemetery, Springfield, Missouri. He lived longer than any of David Murray's sons.

The second son of David Murray, Zelotus, Z.G., or Lotz as he was commonly called, married Maggie Rosenberger on 20 December 1883. In May 1885, Z.G. sought to file for partition of the real estate of David Murray. The minor heirs, Andrew J., William, Evalina Murray and Lawson Banfield were defendants in the case. The Greene County Circuit Court decided that the Sheriff of Greene County sell the real estate at public sale on the court house steps. On 14 November 1885, the sheriff held the sale and the highest bidders were Zelotus G. Murray and Jacob Rosenberger, Zelotus' father-in-law. This sale transferred the Murray farm to Zelotus.

Maggie Rosenberger Murray died 29 July 1891, leaving three young children: Susan, Evalina and Luther. 20 January 1892, Z.G. married Lena Brune and had three more children: Walter, Dorsey and Mamie.

The 1900 Greene County census lists: Zelotus G. Murray, age 39, farmer, Lena C. his wife, age 31, born Germany, Susan, daughter, age 17, Evalina, daughter, age 15, Luther, son, age 11, Walter, son age 6, Dorsey, son age 4 and Mamie, daughter, age 1.

On 13 January 1913, Zelotus G. Murray died intestate. There were letters in his private papers from family members inquiring about his health, but apparently he did not feel that his illness was life threatening. The Missouri death certificate listed his cause of death as La Grippe complicated by pneumonia, with contributory cause of fatty degeneration.

The inventory of his estate showed quite a list of notes payable to Z.G. at 6 percent to 8 percent interest for a total of $22,349. He also had 15 horses and mules, 16 cows and calves, 25 steers, 23 hogs and 35 sheep. His real estate was the land that he inherited and purchased from his siblings. (11) Z.G. was buried at Greenlawn Cemetery, Springfield, Missouri, as was his wife, Lena.

The third son of David Murray, Andrew Jackson Murray, married Lucy Jane Carter 1 September 1886. She was the daughter of Hazen Carter and Elizabeth Banfield Carter and the granddaughter of Qualls and Lucy Banfield.

Andrew was listed as guardian of young Lawson Banfield in 1889, but asked to be discharged from that duty in 1890.

On 20 November 1886, Andy J. and Lucy purchased their first piece of property located East ½ of Lot #1 of NE full quarter of Lot #2, Township 29, Range 22. Andy purchased many pieces of property during his lifetime and was known as a cattle farmer, driving cattle from Arkansas to Greene County and fattening them on his farms. Andy's grandson remembered driving the cattle down the current Glenstone Avenue in Springfield and on to the farm. When Andy died in 1934, his estate inventory had a lengthy list of real estate, many loans to individuals, cash on hand and in the bank and stock certificates in the local stockyards. (12)

In the 1900 Greene County census Andy Murray is listed as age 34, farmer, Lucy J. wife age 33, Ollie, daughter age 12, Edna, daughter age 10, Carl, son age 6, Jettie, daughter age 1.

Andy died 30 July 1934, verified by Missouri Death Certificate. Immediate cause of death was listed as broncho-pneumonia, caused by fractured ribs after being thrown from a horse on his farm. He and his wife, Lucy Jane, are buried at Greenlawn Cemetery, Springfield, Missouri.

The youngest son of David Murray was William Penn Murray. He married Sarah

Jane Stivers, 18 February 1891. He was named earlier buying items from the Louis Dickens estate. He also rented farmland from Aulsie Banfield.

In the 1910 Greene County census, William and Jennie were listed in Franklin Township, just east of Robberson Township. William was 42, farmer, Jennie was 40, Maisie, daughter age 18, Norman, son age 14. A child, Ralph, born and died in 1893, was buried in the Murray Family cemetery.

William P. Murray died 20 March 1911, intestate. He, too, must have not felt his illness was life threatening. In his estate papers, there is one doctor bill for treatment in February and March, but no other information about his illness. There is no Missouri death certificate for William, even though Missouri began recording deaths in 1910. Descendants of William have heard no stories regarding the cause of his death.

Estate papers indicated that William was following the paths of his older brothers and his father. (13) They believed that a successful man owned land, raised livestock and loaned money. William had a considerable amount of money on loan to neighbors and relatives and owned horses and cattle as well as 120 acres in Section 31, Township 30, Range 21. He had pastures rented for his cattle, including land in Taney County, Missouri. His wife made payments from the estate to J.M. McQuerter for care of the cattle in Taney County.

William P., and his wife, Jennie, are buried at Greenlawn Cemetery, Springfield, Missouri.

Evaline, the youngest child and David and Huldah's only daughter, was enumerated in the 1880 census and not found again until the 1900 census. She had married John Makley and was living in Cleburne City, Texas. On the census, John was a mechanic, age 36, born in Ohio, Evaline was age 29, born in Ohio, Marie, daughter was age 7, born in Missouri.

In the 1910 census of the city of Dallas, Texas, Evaline Makley was 36, born Ohio, her daughter Marie was 16, born Missouri. John is not listed in the household and it is believed that they were divorced by this time. In later census reports, Evaline is listed as widow. She was working in Dallas as a stenographer for a book publisher in 1930.

Although Evaline (Evalina, Evalena, Evalene) says she was born in Ohio in the 1900 and 1910 census reports, her correct birthplace was Missouri. Her Texas death certificate states her birthday as 4 November 1873, but she was six months old on the 1870 Missouri census. She died 30 October 1944 and was buried in the Grove Hill Cemetery in Dallas, Texas.

To complete the story, it is necessary to include David Murray's three children by his first wife, Susannah Long. The oldest child, Jerema, was born 8 April 1850 in Wyandot County, Ohio. She married Hamilton Sterling and stayed in Ohio. Jerema and Hamilton had two children: George and Ethel. Hamilton is the son-in-law mentioned at the beginning of this chapter, who visited Greene County and didn't like the land because of all the rocks. Hamilton was originally a farmer, but apparently didn't care for farming because in the 1880 census of Wyandot County his occupation was listed as hotelkeeper. Hamilton, Jerema, and their children, George and Ethel were all buried in the McComb Union Cemetery, Hancock County, Ohio.

Jerema's sister, Ezenith came to Missouri with David and Huldah. On 8 December 1869, she married Rolen Sell, also from Ohio. They had 12 children, including two sets of twins and lived most of their lives in Polk County, Missouri. An extensive history of the Sell family is found in *History & Families, Polk County, Missouri*. (14) Ezenith and Rolen are buried in Oakville Cemetery, Morrisville, Missouri, property which they gave to Oakville Baptist church for use as a cemetery.

Not much is known about David's son, Isaiah. The family story is that he left home at fifteen after his stepmother Huldah tried to whip him with a black snake whip. David's brother, Z.G. noted in his letters when he had received letters from Isaiah and Isaiah wrote occasionally to his sister, Ezenith. The only correspondence from Isaiah in David's papers is the receipt where he acknowledged receiving $100 from his father's estate. It was dated 28 October 1875.

Isaiah was found in the California Voter Register 1866–1898 living in Tulare County. He stated his age as 24, born Ohio, farmer, living in Mountain View, Tulare County, California. The date of his voter registration was 3 August 1875. Isaiah died sometime after 1875.

The David Murray Farm Today

The original 320 acres which David Murray bought has been reduced in size by approximately 100 acres due to a number of reasons: the impoundment of water on Little Sac River and the creation of McDaniel Lake, a source of water for the city of Springfield, the change in the highway in 1928 and the construction of the new Highway 13 in 1975. In 2008 and 2010 twenty eight acres were sold and gifted to the Springfield-Greene County Park Board to provide a hiking and biking trail. The trailhead was named the David C. Murray Trailhead. The farm has been recognized as a Century Farm of Missouri because of its continuous ownership by one family for over 100 years.

References

(1) *Wyandot Democratic Union*, (Upper Sandusky, Ohio).

(2) Private papers David C. Murray.

(3) Greene County Missouri Court Case Book G, 278.

(4) Greene County Missouri Court Case Book G, 306.

(5) Greene County Missouri Circuit Court Case File #3834.

(6) Greene County Missouri Circuit Court Book, 217.

(7) Greene County Missouri Probate File #7460.

(8) *Canada Farmer Magazine*, 15 April 1865, "Building Styles," OntarioArchitecture.com.

(9) Greene County Missouri Deed Book 50, 345

(10) Greene County Missouri Deed Book 50, 345, acknowledgement by notary

(11) Greene County Missouri Probate File #7732.

(12) Greene County Missouri Probate File #11306.

(13) Greene County Missouri Probate File #7646.

(14) *History of Families, Polk County, Missouri*, (Paducah, Kentucky: Turner Publishing Company), 328.

Chapter 7

Burials in the Murray Cemetery

The cemetery is located in Section 35, Range 22, Township 30, on the hill overlooking the historic Murray farmhouse. An inventory was completed in 1985 by Bettie Hickman and Mary Cunningham, members of the Ozarks Genealogical Society and it can be found in Volume III, Greene County Missouri cemeteries. In addition to the tombstones with inscriptions, they note at least 45 fieldstone markers. This cemetery should not be confused with the Murray Cemetery located in Murray Township, which is just north of Willard, Missouri. No relationship has been found between the Murray families of Robberson Township and Murray Township.

Hickman and Cunningham state that the cemetery was overgrown with weeds and stones were broken and in disarray. That is still true. Maintaining an old cemetery is a challenge for the farm owner. If the cemetery is fenced, grass and weeds grow and the area requires hand mowing. If the cemetery is unfenced, the cattle rub against the stones, knock them over and push stones into the grass with their hooves; they do, however, mow the grass. The cemetery is now fenced, protecting the stones from cattle, but allowing the fescue grass to grow thick, and in some cases, the fescue obscures the fieldstone markers.

When I was a child, I remember looking at all the fieldstones marking gravesites and wondering who those people were and why they were buried in our family cemetery. One plot was completely enclosed with fieldstones, plus head and footstones. There is no historical information about who is buried there. A few graves were blessed with formally inscribed tombstones.

Because there were so many fieldstones in the cemetery, neighbors must have buried their family members there between 1830 and 1870. Tombstones may have been unavailable, or if they were, the cost of purchasing and transporting the stone five miles north of Springfield was prohibitive. Some of the burials could be family members of the early settlers in Sections 26 and 35.

I knew who the Murray relatives were that were buried in the cemetery and have found newspaper obituaries about some of the other people who were buried without gravestone markers. All the individuals identified who had stones were relatives or neighbors.

Murray Family

Two arched stones with clasped hands mark the graves of David C. Murray and his wife Huldah M. Murray. David was born 16 October 1822 and died 9 April 1872. Huldah was born 22 September 1839 and died 5 July 1879. They were the parents who brought their children to Greene County in 1867.

A large upright carved granite stone marks the grave of Margaret Murray, the first wife of Zelotus Murray, son of David and Huldah. She was born 15 August 1862 and died 29 July 1891. It is believed she died at childbirth or shortly after.

David Murray tombstone.

Margaret Murray tombstone.

Next to her stone is one for an unnamed infant who died August 1891.

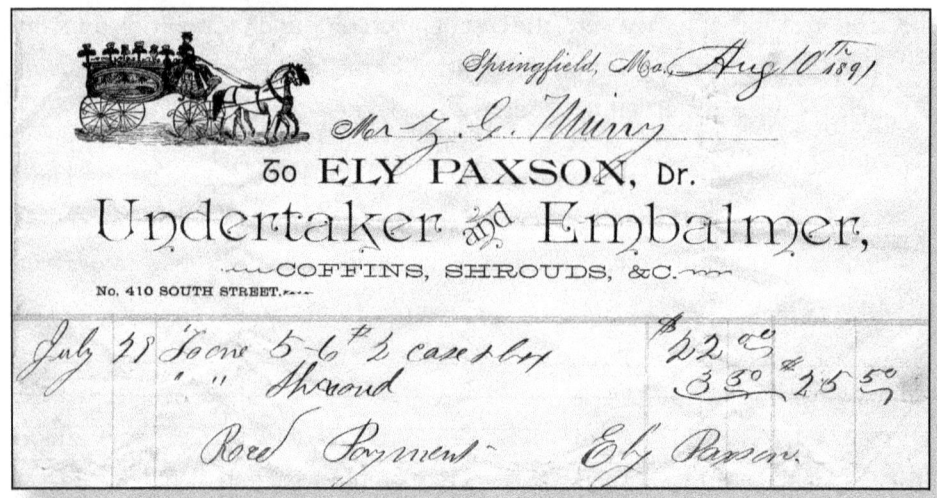

The bill for Margaret's coffin from Ely Paxson, Undertaker and Embalmer, 10 August 1891

Zelotus' and Margaret's son Freddie is to the north of Margaret's stone. He was born 29 June 1887 and died 21 December 1887. The family story is that one of his older siblings was holding him and somehow he fell into the fireplace and was severely burned, resulting in his death.

Apparently Z. G. Murray ordered gravestones for David and Huldah Murray and little Freddie in February 1890. The order placed for the stones was found in the Murray family papers. The pair of stones from Italian marble, with clasped hands, were for David and Huldah. The white stone with lamb engraved on the front was for the son, Freddie.

Freddie Murray tombstone

Order for tombstones from Buffalo Marble Works, February 1890

These documents show that there were no permanent markers in the Murray Cemetery until 1890 when Zelotus marked the gravesite of his son and the graves of his parents.

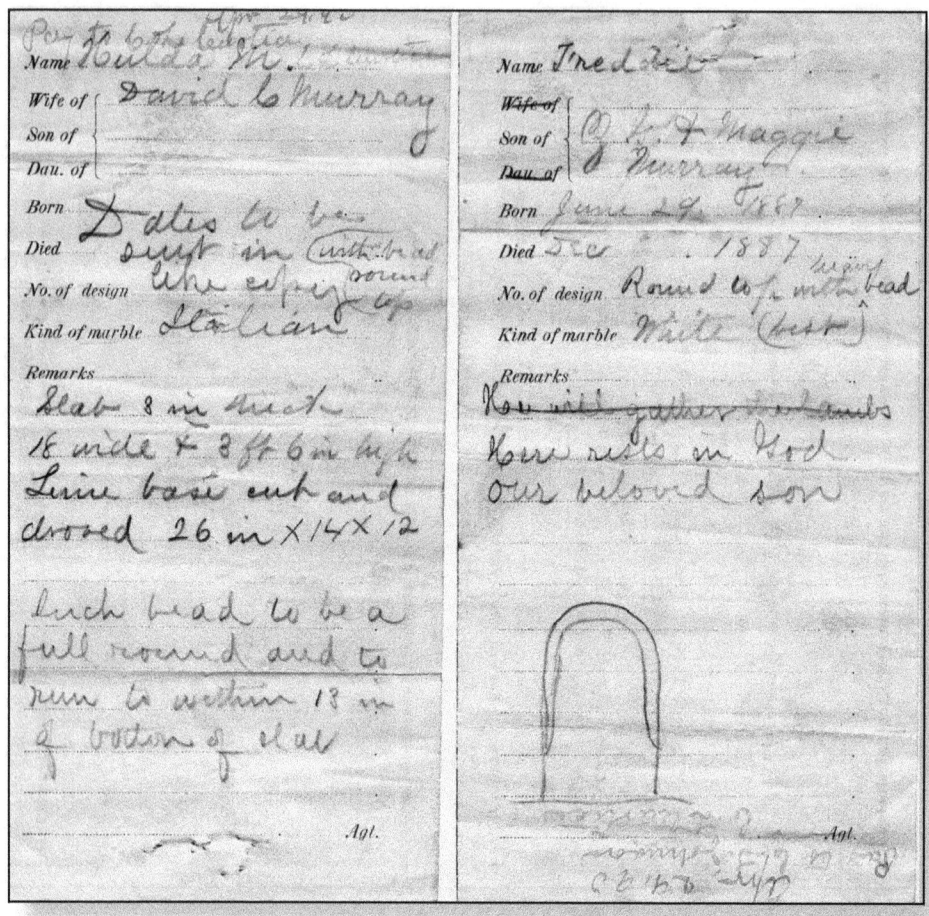

Description of tombstones ordered 1890

The following bill, found in Murray family papers, indicates that in February 1888, Zelotus Murray had workers clear the grounds and fence the cemetery. Perhaps Maggie requested this after the death of Freddie. From the number of hands used, the graveyard must have been badly overgrown. The workers were brothers Z.G., Andy and William Murray as well as William Jones, Louis H. Dickens and James M. Dickens who were later buried in the cemetery.

Bill for workers to clear the cemetery >

murray the is ti for & san

6	February the 6 day 1888		
7	Wire and Staple	4	70
8	Work on clearing Grave		
9	yard 11 hands one day	8	25
	fore hands and one half	3	57
	fore hands one day	3	00
	one hand one half day		37
	Tos hands one day	1	50
10	Sire hands one half day	4	85
	lumber for gate		40
	Post	5	00

the the is ti mingie mamin
man is ph
3 I G murray
maggie murray
Sucy murray
    murray
1 Kate pluelaeer

The other young Murray was Ralph, born 15 January 1893 and died 16 October 1893. Cause of death unknown. Ralph was the son of William P and Jennie Murray. William was a brother to Zelotus, son of David and Huldah. The stone has a broken tree trunk with a lamb, signifying premature death and innocence.

Another family connection is noted in the row behind Margaret Murray's stone. There is a dark gray granite stone with the surname Brune. Zelotus' second wife was Lena Brune and this stone marked the graves of her mother and sister. Lena's mother, Willhmina was born 6 February 1847 and died 14 January 1899. Racie, Lena's sister, was born 31 July 1881 and died 24 December 1899. The Brune family had immigrated to the United States in 1872 from Germany.

Stone for Willhmina and Racie Brune

On Tuesday evening, 26 December 1899, The *Springfield Leader Democrat* announced the death of Miss Racie Brune age 18, who died Sunday of typhoid fever and was buried in Murray Cemetery. On December 30, they announced her death again and said she was engaged to be married to a man whose last name was Thomas.

Louis Dickens tombstone

Dickens Family

Individuals with the Dickens surname were neighbors, but they did not intermarry with members of the Murray family. Perhaps they requested burial in the Murray Cemetery because of the close proximity to their homes.

A moderately large obelisk stone marks the grave of Louis H. Dickens, born 3 July 1840 and died 19 January 1902. It has the additional inscription, "Born in Bedford County, Tennessee." At the time of his death, Louis owned the S ½ of the SW ¼

S25, R22, T30, which is the next section east of Section 26. See history of the Dickens Family.

There is a square stone, off base, marking the grave of Henry G. Dickens, born 3 September 1866, died 17 December 1895. He is apparently the son of Stephen Dickens because he is listed in the Stephen Dickens family in the 1870 and 1880 Greene County census.

Another grave marked by a stone is that of Luther M. Dickens, born 30 November 1892, died 21 August 1894. It is not known to which family unit he belongs. However there was an announcement on 22 August 1894 in the *Springfield Republican*, "Infant son of William Dickens died yesterday and will be buried in the country." This date would coincide with the date on the Luther M. Dickens tombstone. Since the names Willis and William were frequently used interchangeably, this could be the son of Willis Dickens.

Henry A. Dickens was reported buried in the Murray Cemetery by the *Springfield Republican*, 22 June 1900. He age was 36 and he was the son of Louis H. Dickens. There is no stone marking his grave, which is surprising since his father was still living.

James M. Dickens, brother to Louis H., was reported buried in Murray Cemetery, in the *Springfield Republican* on 10 October 1908. He died 9 October 1908, at the age of 64. No stone marking his grave.

Sarah Melissa (Austin) Dickens, wife of James M. Dickens, was buried in Murray Cemetery. She was born 29 May 1848 and died 10 November 1938. Date of death and burial information was confirmed by Missouri Death Certificate #29238. Her grave is unmarked. The *Springfield Daily News*, 11 November 1938, stated that she was age 90 and died at her home on Route 5, Springfield. Survivors were her son, Sherman, daughters Anna Bire (Vire) and Eva Bire (Vire), 6 grandchildren and 14 great grandchildren. The funeral service was conducted at Glidewell Baptist Church, burial location not given.

The *Springfield Leader Democrat*, 23 March 1900, reported the death of Mrs. Katherine Austin, age 91, yesterday at the home of J.M. Dickens, six miles north of the city, buried in Murray Cemetery. There is no marker for her grave. See information under Austin Family in Chapter Four.

After Zelotus Murray died in 1913, no other person was buried in the Murray Cemetery with the exception of Sarah Melissa (Austin) Dickens, wife of James M. Dickens. Zelotus' son, Dorsey, gave permission for the burial so that she could be buried where her husband was buried.

Maryan Martha and William Jones
I have found no family relationship between the Murray family and the Jones family. Perhaps because they were neighbors, they requested to be buried in the Murray Cemetery. Their stones were recorded in 1985 and have since been covered by fescue grass. The stones read: "Maryan Martha Jones, died 29 May 1907, ae 84 years" and "William G. Jones, 16 Dec 1826—5 Feb 1908."

William and Martha Jones were living in North Carolina at the time of the 1850 census. The Surrey County, North Carolina census listed William Jones, age 21, laborer, born Virginia, Martha age 24, born Virginia and Madison, age 1, born Virginia. In Virginia Marriages 1740-1850, there was a marriage listed for William Jones and Martha Chaney, 15 January 1845, Pittsylvania, Virginia.

In the 1860 census of Robberson Township, Greene County, Missouri, Wm Jones was listed as age 33, farmer, $1500 value real estate, $2200 personal property, born Virginia, Martha age 32, born Virginia, Madison, age 13, born North Carolina, Henry age 5 and Mildred age 2, both born North Carolina, Lucy Jones, age 65, personal property value $2000, born Virginia.

In the 1870 census of Robberson Township, Greene County, Missouri, William G. Jones was listed as age 43, farmer, real estate value $2000, personal property value $400, born Virginia, Martha age 40, born Virginia, Madison age 23, born North Carolina, Henry G. age 15, born North Carolina, Sarah Chandler age 37, born Virginia, Lucy Jones age 78, blind, born Virginia. They were living next to Stephen and Henry Chandler. In the 1876 plat book of Greene County, William Jones property is adjacent to Huldah Murray and Henry Chandler.

Solomon and Milly Cotner

There is no relationship between Solomon Cotner (Cottner) and the Murray family, although after considerable research, a relationship was found between the Cotner and Dickens families.

Solomon was an early settler in Greene County and noted by Holcombe as "being one of the early settlers, who, with Jacob Painter, could kill more game, and they were considered the most expert hunters in the country, and long after wild game had disappeared, they could find and kill deer almost in sight of town, when no one else could." (1)

In the 1840 census of Greene County, Missouri, he was listed as Solomon Catner, male 40-50, 1 female 5-10 and 1 female 30-40. Other names appearing on the page were early and well-known settlers, John P. Campbell, D.D. Berry, Peter Apperson and Joseph Rountree.

From the probate file of Thomas James, there is a small account book containing dates and names of farmers who had their mares serviced by James' jack. (2) On 26 May 1837, there is the entry: "Sollomon Curtner one sorrel mare to the Jack by ensurence part of Kindals Class." It is not known what the reference to "ensurence" or "Kindals Class" means, but both terms were used frequently in Mr. James' journal.

In the 1835 Greene County assessor's list, Solomon is assessed for 1 horse, value $25, 1 poll tax. There is also a Daniel Cotner assessed for 1 horse valued at $40, 2 cows valued at $20, no poll tax noted.

In the 1843 Greene County assessor's list, Solomon is assessed for two horses valued $60, 2 cattle valued $12. Daniel Cotner is assessed and listed as deceased and Washington Cotner is assessed one poll. No relationship had been found between these men until an entry on Ancestry.com provided a clue.

A family tree on Ancestry.com listed a David C. Cotner who was born 1776 in North Carolina and who had sons: Solomon born 1800, Daniel born 1805, Peter born 1809, George Washington born 1814, John born 1816, Amziah born 1825, William born 1828. Solomon, Daniel and George Washington apparently were in Greene County in 1843, as indicated by the early tax assessments.

Daniel Cotner was listed in the 1840 census of Greene County: 1 male under 5, 1 male 20-30, 1 male 30-40, 2 females under 5 and 1 female 20-30. He applied for a land patent from the U.S. government on S6, T30, R24. Patent issued in 1845. This land is at the western edge of Greene County, located next to Dade County. Daniel died in 1843 and Solomon was appointed executor of the estate. An inventory of his estate showed animals and accounts of $229, but no value given to the real estate. In the Greene County Probate Book A, (3) the court decided that the amount of the estate and the amount due the widow was too small for administration. The estate was delivered to the widow. Solomon had indicated on his bond information that there were three heirs, but nothing has been found about the heirs of Daniel Cotner.

Solomon and Daniel's father, David C. Cotner, applied for a land grant in S14, T30, R25, Dade County. Patents issued in 1850 and 1853. David was listed in the 1850 census as a farmer, age 76, real estate valued at $600, born in North Carolina. His wife was Milly, age 74. David died in 1850 and Milly died in 1857, both buried in Sinking Creek Cemetery in Dade County, Missouri. (4)

George Washington, probably the Washington Cotner assessed in the 1843 Greene County tax records, or Uncle Wash, as he was called in Dade County, bought 40 acres from his father, land which had been obtained by David's land patent. George W. purchased several other patents himself. Patent # MW-0634-005 was for his service in the Mexican War when he served as 1st Lieutenant, Captain J.J. Clarkson's Company F. He enrolled 26 April 1847 at Greenfield, Missouri and was honorably discharged 15 October 1848 at Independence, Missouri. (5)

Meanwhile Solomon in Greene County had increased his fortunes. In 1851 he was assessed for one slave valued at $330, 2 horses valued $50, 5 cattle valued $25, money on loan $110 and real estate, S6, T29, R20, 83 acres valued $100. The 1850 slave schedule census listed Sol Cotner with one slave, a black female, age 9 years.

By 1856 Solomon was assessed for two slaves valued at $800, 2 horses valued $90, 2 cattle valued $20 and the same real estate as listed in 1851, valued at $800. In the 1860 slave schedule census, Campbell Township, a "G" or "S" Catner is listed as owning three slaves. The schedule listed 1 female age 19, black, 1 female age 6, black, 1 male age 2, black.

The 1860 census, Campbell Township, Greene County, Missouri, listed Solomon as age 60, farmer, real estate valued at $2000, personal property valued at $3000. Both he and Milly, age 56, stated they were born in North Carolina.

The *Springfield Weekly Patriot*, during 1870, announced the death of Milly Cotner. "Died November 3, 1869, Mrs. Milly Cotner, wife of Mr. Sol Cotner, an old citizen of this county, aged about sixty eight years. Omitted from Mortality Schedule."

In the 1870 census of Campbell Township, Greene County, Missouri, Solomon, now a widower, was listed as age 70, farmer, real estate valued at $3200, personal property $700. Living with him were Geo Vaugh, age 49, farm hand, born Alabama, Maria, age 34, keeping house, born Tennessee, Amanda, age 15, domestic servant, born Missouri, and the other children, Joseph age 12, Edward age 9, Eliza age 6, Silas age 4, all born in Missouri. George, Maria, Edward, Eliza and Silas are identified as black; Amanda and Joseph are identified as mulatto. No connection was made with this family until Solomon's will was found.

On 11 December 1871, Solomon married Mrs. Elizabeth Frankum in Dade County, Missouri and moved there. (6) He is listed in the 1880 census as Solomon age 80, farmer, Elizabeth age 49, wife, John Frankum age 17, stepson, Bertha Duff, age 3, grandchild. Elizabeth died 3 March 1884.

Why did Solomon move to Dade County after he had spent most of his life in Greene County, and what happened to the real estate he owned in Greene County? A trip to the courthouse in Dade County, specifically the probate office and a search of deeds in the Greene County Archives answered these two questions.

He must have moved to Dade County because his relatives were there, George Washington and family and the family of his sister Mary J. Cotner Patterson. Mary J. Cotner was not listed in the Ancestry.com family tree as a descendant of David and Milly Cotner. However an article in *History of Dade County and Her People*, stated that, "Millie Cotner Patterson and her husband, William H. Patterson migrated to Dade County in 1842, coming with four children, and an uncle of Mrs. Patterson's, Uncle Warsh Cotner." (7) (Uncle Warsh is probably George Washington Cotner and would be a brother to Mrs. Patterson).

First, the property: On 1 March 1880, Solomon and Elizabeth Cotner sold the north fractional half of the SW ¼ in S6, T29, R21, in Greene County, Missouri containing 42 acres to Maria Vaughan. (8) This is the same Maria as in the 1870 census. The name Vaugh on the census was apparently an error. On Maria's death, the property was to pass to four of her children: Edward Cotner, Tyler Cotner, Cora Vaughan and Eliza Cotner. On 11 March 1880, Solomon and Elizabeth sold to Manda Ann Smith lot #1 in the NW full quarter S6, T29, R21, consisting of 39 and 91/100 acres m/l. (9) This is the same person as Amanda, listed in the 1870 census, who married Taylor Smith, 30 July 1871. (10)

The probate file in Dade County revealed the answer to the questions about Solomon's relatives and why his wife Milly was buried in Murray Cemetery. (11) His will, signed 10 September 1891, directed the following: (1) payment of debts and funeral expenses (2) bequeath to my first wife's nephews, Lewis and Madison Dickens, the sum of five hundred dollars to be divided among them and there [sic] three brothers if they be living yet, (3) bequeath to Joseph Cotner the sum of three hundred dollars (4) bequeath to Amand Smith fifty dollars (5) bequeath to my beloved sister Mary Patterson and my beloved brothers Peter and Daniel Cotner and their bodily heirs the remainder of all my estate to be divided equally among them, (6) have already given to G.W. Cotner as much as will be left for the other heirs (7) I nominate and appoint my friend W.T. Hastings to be executor.

Solomon had an estate with considerable assets. He owned 80 acres in S11, T31, R27 in Dade County, Missouri, plus certificates of deposit and loans to persons bearing 10 percent interest.

In the first settlement of the estate, William Hastings, executor, paid $100 each to Solomon Dickens, James M. Dickens, Lewis Dickens, S. F. Dickens, W.R. Dickens, sons of William B. Dickens. This explains why Milly—Amelia Dickens Cotner — was buried in the Murray Cemetery. She was buried in a cemetery close to her Dickens family. Louis H. and James M. were later buried there.

By naming "my beloved sister Mary Patterson," and brother's Peter and Daniel, Solomon confirmed the information found on Ancestry.com regarding the family configuration.

Solomon had no surviving children. The female listed in the 1840 census apparently died young and there is no information where she was buried. Could she have been buried in the Murray Cemetery with one of the unknown fieldstones identifying her grave? Since he had no children, did Solomon feel very close to Maria Vaughan and her family? He provided for all of them who were in his household in 1870 either before his death or in his will. By selling the 42 acres of his land in Campbell Township, he provided for Maria during her lifetime. The property was to pass to her children, Edward, Tyler, Cora Vaughan and Eliza, at her death. (Could the Tyler named in the deed and the Silas named in the census be the same boy?) Then Solomon sold 39 ½ acres of his land in Campbell Township to Amanda Smith and gave her $50 in his will. The other child, Joseph Cotner was given $300.

The final answer to this puzzle came when Amanda Smith's death certificate was found. Missouri death certificate #17539 identified her as female, black, married, born 12 June 1853, died 23 May 1911. The informant was her husband, Taylor. He gave the name of her father as Solomon Cotner and her mother, Maria Vaughan. Maria must have been the little slave girl listed in the 1850 census. The persons in the 1860 census must have been: Maria age 19, Amanda age 6 and Joseph age 2. Amanda and her husband, Taylor, are both buried in Hazelwood Cemetery, Springfield, Missouri. See further research on this family at the Greene County Archives.

There was one final interesting note in Solomon's probate file. The executor paid $8.35 for "railroad fare for body and mourning friends to Springfield." Paxon's hearse to the cemetery, a distance of six miles, cost $10. Mark Gault was paid $10 for providing three carriages. One can imagine the funeral procession, by train from Greenfield to Springfield, the casket and mourners on the train and then the carriage procession carrying the mourners to the cemetery.

Solomon was highly esteemed in Greene and Dade counties and frequently referred to as the oldest man living in the area.

The *Greenfield Vedette* reported on October 15, 1891: "Died- At the residence of W. T. Hastings, Esq. near Everton, Mo. At 1:30 a.m. on Thursday, Oct 8th, Solomon Cotner, aged 91 years 8 months and 10 days. Mr. Cotner resided near Greenfield for a number of years, and during all that time was a familiar figure on our streets. Two years ago he took up his abode with Esq. Hastings, near Everton, since which time he has visited in the county seat only at long intervals. He was born January

29, 1800. We are not positive as to the State of his nativity; but think it was North Carolina. Owing to failing eyesight and other inevitable infirmities of Old age, he has since his removal to Everton, gone but seldom from Home. On Friday October 9th, his remains, in compliance with his last request, were committed to a grave in the Murray burying ground 3 miles from Springfield, beside the grave of his first wife. May he rest in peace." The gravestones for Milly and Solomon must have been placed in the cemetery in 1891, when Solomon died. Because all of the stones with inscriptions were placed after 1890, it would have been unusual for Milly to have an engraved stone placed in 1870, when all the other stones were fieldstones.

All the others buried in the Murray Cemetery remain unknown unless a researcher discovers a newspaper obituary or a descendant can verify that an ancestor was buried there. It must have been used as a neighborhood cemetery long before David Murray purchased the farm. We can only wonder who those people were.

References

(1) Return I. Holcombe. *History of Greene County, Missouri*, (St. Louis, 1883) 150.

(2) Greene County Missouri Probate File #5018.

(3) Greene County Missouri Probate Book A, May term 1843.

(4) Personal papers, Cotner Family, at Dade County Genealogical Library, Greenfield, Missouri.

(5) "Soldiers Records of War 1812-World War I," Missouri Secretary of State digital collection. www.sos.mo.gov.

(6) Personal papers, Cotner Family, at Dade County Genealogical Library, Greenfield, Missouri.

(7) R. Ludwig and A.J. Young, *History of Dade County and Her People*, (Greenfield, Missouri: Pioneer Historical Company, 1917), 220.

(8) Greene County Missouri Deed Book 38, p. 458.

(9) Greene County Missouri Deed Book 38, p. 457.

(10) Greene County Marriage Book C, p. 358.

(11) Solomon Cotner Probate file #976. Dade County, Missouri.

Epilogue

The Boonville Road, which crossed the farm between Little Sac and Little Dry Sac, is now a private lane in front of Huldah Murray's house. Because research has shown that Boonville Road was once used by the Butterfield Overland Mail, the lane has been designated a Greene County Historic Site.

Butterfield Route on Murray Farm

In 2008 and 2010, the Springfield-Greene County Park Board acquired 28 acres, located on the south side of Little Dry Sac River, from David Murray's great granddaughter and her husband. The Ozark Greenways, a part of the Springfield-Greene County Parks, opened The South Dry Sac Greenway, a hiking and biking trail, which can be accessed at the David C. Murray Trailhead on Farm Road 141. That location may be very close to one of the fords noted in David Murray's map of 1867 (page 23). Kiosks have been placed beside the hiking trail to interpret the history of the Boonville Road, the Butterfield Mail and the David Murray family.

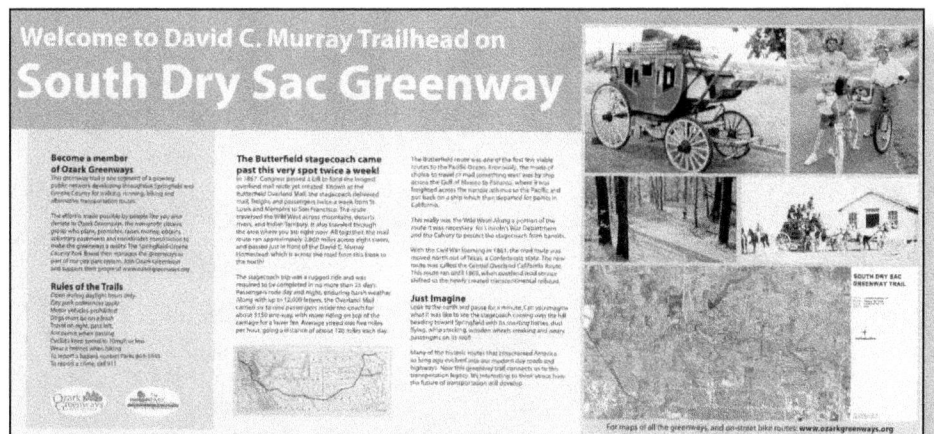

Kiosk #1 at David C. Murray Trailhead on South Dry Sac Greenway

Butterfield Overland Mail Route

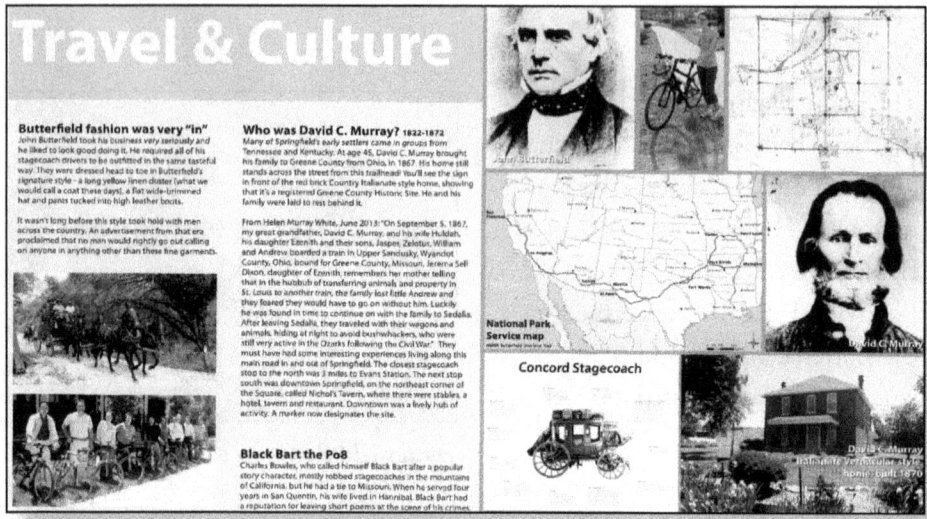

Kiosk #2.

Future plans are for the trail to continue east to Lost Hill Park, the home of the Banfield family described in Chapter Five. The great grandson of Qualls Banfield and grandson of Andy J. Murray sold the Banfield farm, known as the Owen Farm, to the Springfield-Greene County Park Board. The continuation of the South Dry Sac Greenway Trail will once again connect the Murray and Banfield families.

Huldah's Murray's House. Photo taken 2013.

Huldah's house, 135 years old in 2013, was placed on the Greene County Historic Sites Register in 1988 because of its Italianate design, a type of architecture unusual for a country home. Although the interior has been remodeled and updated, there have been few exterior changes since it was built. The house, used by family and friends, is not open to the public.

A

Adams, Archibald Clinton, 25, 26
Adams, Frances McClure Dryden, 26
An Index of the Springfield Land Office Sales Book, 1833-1892, 25
Apperson, Peter, 115
Austin, Anna Elizabeth, 62, 67
Austin, Calvin, 65
Austin, Cate (Catherine, Katherine), 65, 66, 68, 113-114
Austin, Claudy, 68
Austin, Eliza A., 66, 68 (wife of John)
Austin, Elizabeth Ann, 67
    (daughter of John)
Austin, Elizabeth Calvert (Halbert), 64, 65
Austin, Elizabeth Jane, 65, 66-67
    (daughter of Samuel)
Austin, Francis M., 68
Austin, Jane, 66 (daughter of Samuel)
Austin, John, 63, 65, 66, 67, 68, 97
    brother of Samuel III
Austin, John Calvin, 65, 66, 67
    (son of Samuel III)
Austin, Margaret A., 68
Austin, Samuel III, 66, 68, 97
    land patents, 63
Austin, Samuel II (Jr.), 64, 68
    (father of Samuel III)
Austin, Sarah (Sally, Sallie), 68, 69
    (sister of Samuel III)
Austin, Sarah A.M., 65, 66
    (daughter of Samuel III)
Austin, Sophronia, 67

B

Baker, Silas, 31, 48, 52, 55
Banfield, Aulsie (Alice) McGrew (McGraw), 79, 85, 103
Banfield, Allie-86,
    (wife of Robert Lawson Banfield)
Banfield, Bernice, 86, 87
Banfield, Bradley, 80
Banfield, Columbus, 81
Banfield, Elizabeth, 80 (sister to Qualls)
Banfield, Elizabeth F., 80, 81
    (daughter of Qualls)
Banfield, Elizabeth Carter, 79
    (wife of James Banfield, sister-in-law to Qualls)
Banfield, Fanny (Fannie) (slave), 82-83, 85
Banfield, George, 80
Banfield, Gladys, 86
Banfield, Huldah Murray, 66, 79, 83-84
    See Murray, Huldah
    ante-nuptial agreement, 83, 84
    house, 98-99, 123
Banfield, James, 79 (brother of Qualls)
Banfield, James A., 82
    (son of John and Mary Banfield)
Banfield, James Wilson, 85
Banfield, John, 78, 79 (father of Qualls)
Banfield, John, 55, 71, 78, 80, 81-82
    (brother of Qualls)
Banfield, John (son of F.T. [R.Q.]), 81, 85
Banfield, John (son of John), 82
Banfield, Lawson C., 85-87, 101, 102
Banfield, Lewis H., 73-74, 80, 83
    (son of Qualls)
Banfield, Lobetta A., 87
Banfield, Louisa, 81, 82
Banfield, Lucy Ann Warren, 78, 79, 80, 81, 82, 83, 102 (wife of Qualls)
Banfield, Mary E. Dickens, 71, 80, 81, 82
    (wife of John Banfield, son of John)
Banfield, Mary J., 80 (daughter of Qualls)
Banfield, Maude, 87
Banfield, Nancy, 81 (daughter of Qualls)
Banfield, Nancy Giboney, 83
Banfield, Nancy J., 82
    (daughter of John and Mary)

Banfield, Quarles, 81
  (son of R.Q. Banfield)
Banfield, Rachel, 82
Banfield, R.L. (Qualls), 66, 81
Banfield, Roger Lawson, 83, 84, 85–87
Banfield, Roger Qualls, (Quarles), 55, 66, 78–79, 80–84, 102
  death of Huldah, 84
  suit over horse, 58, 94–95
  will, 85–86
Banfield, Tabitha, (Tabithy) 71, 78–79
  (wife of John)
Banfield, Urban, 86, 87
Banfield, Valls (Qualls), 81
Banfield, Westley, 81
Banfield, William B., 80
  (probable brother of Qualls)
Banfield (Bandfield) Tabitha 79–80
  (sister of Qualls, married to James Hagen)
Banfield (Bunfield), John, 71 (father)
Banfield (Bunfield), John (father), 71
Banfield (Bunfield), Mary Dickens, 71
Banfield (Bunfield), Tabitha, 71
Bath, Mary C., 72
Battle of Pea Ridge, 74
Battle of Springfield, 16–17, 34
Battle of Wilson's Creek, 60, 74
Benton Thomas Hart (Senator), 2
Berry, D.D., 8–9, 115
Berry, William B., 73
Beyond the Mississippi, 21
Bird, Daniel, 49
Bire (Vire), Anna, 113
Bire (Vire), Eva, 113
Bledsoe, Absolom, 5
Bledsoe, Lewis, 5
Bledsoe's ferry, 5
Bolivar Road, 39, 42
Boone, Nathan, 13
Boonville, Mo., start of Santa Fe trail, 4

Boonville Road
  in 1844 map, 14
  on 1876 map, 41
  becomes Bolivar Road, 39
  as Broadway of Springfield, 91
  in Civil War maps, 16–17
  commercial uses, 8–9
  confusion over, 6–8
  named Highway 13, 30–31
  present-day, 121
  proposed road change 1904, 43
  proposed road changes 1928, 44, 45
  as state road, 15
Breach, Simon, 65
Britt, Randolph, 78
Brune, Racie, 112
Brune, Willhmina, 112
Buffalo Marble Works, 109
Bunfield. *See under* Banfield
Butterfield, John, 20, 38
Butterfield Overland Mail Route
  centennial celebration, 34–35
  southern route ends, 38
  stations on, 21–22
  travelers' reminiscences, 34
Butterfield stage route, 20

C

Campbell, John P., 115
Cannefax and Moore (appraisers), 56
Cannon, Robert, 57–58
Carter, Elizabeth Banfield, 83, 85, 102
Carter, Hazen, 79, 83, 102
Carter, James, 79
Carter, Jeremiah, 79
Carter, Louisa, 78
Carter, Sarah, 79
Cate, Samuel Burlin, 81
Catner. *See under* Cotner
Cedar Bluff Baptist Church, 67
Chandler, Daniel, 48, 61

Chandler, Eliza E., 62–63, 72
Chandler, Elizabeth, 61
Chandler, George, 62
Chandler, Henry R., 48, 61, 62–63, 72, 114
Chandler, James M., 48, 61, 62
Chandler, J.C. (Rev.), 66
Chandler, Jesse, 62
Chandler, John, 61 (son of Daniel)
Chandler, John, 62 (son of Stephen)
Chandler, John M. (Rev.), 62, 67
Chandler, Judy Akins, 62
Chandler, Margarette, 61
Chandler, Mariah, 61
Chandler, Martha J., 62
Chandler, Maude, 62
Chandler, Nancy, 61
Chandler, Sarah, 114
Chandler, Sarah Hall, 62 (wife of Stephen)
Chandler, Stephen, 60–61, 62, 74, 114
Chandler, Viola, 62
Chandler, William P., 61 (listed on 1850 census as son of Daniel)
Chandler, Willis, 61 (listed on 1850 and 1860 census as son of Daniel)
Chrisman, William, 68
Colton's Railroad and County Map of the Southern States, 31–32, 33
Conkling, Roscoe and Margaret, 22–23, 28
Cook, Ezekiel, 31, 81
Cotner servants (Solomon), 117
Cotner, Amziah, 115
Cotner, Daniel, 115–116, 118
Cotner, David C., 115
Cotner, Edward, 118
Cotner, Eliza, 118
Cotner, Elizabeth Frankum, 117
Cotner, George Washington, 115, 116
Cotner, G.W., 118

Cotner, John, 115
Cotner, Joseph, 118, 119
Cotner, Milly (Amelia Dickens), 117, 118, 120
Cotner, Milly 116 (wife of David Cotner)
Cotner, Peter, 115, 118
Cotner, Solomon, 69, 115–118, 119–120
Cotner, Tyler, 118
Cotner, William, 115
Cotner (Cottner, Catner), Solomon, 115
Crockett, Andrew, 56
Crockett, Martha, 56
Crockett, Mary, 56
Crockett, Thomas, 56
Cumberland Road, map, 6

D

Dameron, James A., 53
Dameron, Mary Frances, 53
Daniel, Drucilla Halburt, 58, 64
Daniel (Daniels), John, 57, 58, 73, 74 (son of Tapley)
  suit over horse, 58, 94–95
Daniel, Judy Thornhill, 55
Daniel, Keziah Thornhill, 55
Daniel, Kiziah, 57–58
Daniel, Lucinda, 57, 94
Daniel, Margaret (Crockett), 55–56
Daniel, Nancy (Cain/Kain), 55–56
Daniel, Sarah, 55–56
Daniel, Tapley, 15, 31, 48, 50, 52, 55, 59, 65, 93, 97
  assessments and land patents, 56–57, 63
Daniel, Thomas, 55, 56–57
Daniel, Wiley, 58
Daniel, Willis, 58
David C. Murray Trailhead, 44, 105, 121–122
Davidson, William, 60
Davis, Arthur, 31, 48, 51–52

Davis, David R., 94
Davis, Stephen H., 50
Davis Cemetery, 51
Deeds, Bennett, 80
Deeds, John H., 78
Delaware Indians, 47
Dickens, Andrew R., 70
Dickens, Clinton, 72, 82
Dickens, Daniel, 69
Dickens, Elizabeth, 82
Dickens, George W., 72
Dickens, Henry Allen, 72, 113
Dickens, Henry G., 70, 113
Dickens, Hugh, 69
Dickens, James Henry, 69, 70
    (son of Stephen)
Dickens, James M., 66, 67, 71, 73, 110, 113, 114, 118 (son of William)
Dickens, Jesse, 69
Dickens, John P., 71 (son of William)
Dickens, John Madison, 57–58, 69, 70, 73 (son of Stephen)
Dickens, Joseph H., 69
Dickens, Joshua, 69
Dickens, Josiah, 70
Dickens, Juda (Judy) Elizabeth Scott, 70, 72
Dickens, Lewis, 70, 118
Dickens, Linda, 72, 82
Dickens, Louis (Lewis) Henry, 67, 71–73, 82, 102–103, 110, 112–113
Dickens, Luther M., 113
Dickens, Madison, 70, 118
Dickens, Mary Banfield, 70, 72
    (daughter of William)
Dickens, Nancy, 69, 70, 71
Dickens, Nancy (wife of William), 70
Dickens, Oliver M., 72
Dickens, Polly McDerment, 69
Dickens, Sarah Melissa (Austin), 68, 113, 114
Dickens, Sarah Terrell, 71–72
Dickens, S.F., 118
Dickens, Sherman, 113
Dickens, Solomon J., 69, 70, 73,
    (son of Stephen)
Dickens, Solomon F., 71, 118
    (son of William)
Dickens, Stephen, 58, 69, 71, 113
    land patents, 70
Dickens, Stephen P., 71
Dickens, Stephen (son of William), 70
Dickens, Thomas, 69, 70
Dickens, Viola Chandler, 62
Dickens, Walter, 72
Dickens, William B., 69, 70, 71, 80, 118
Dickens, William F., 69, 73, 74
    (son of Stephen)
Dickens, William R., 71, 118
    (son of William)
Dickens, Willis E. (William), 70, 113
    (son of Stephen)
Dickerson Park Zoo, 35
Dishman, Mary Banfield, 83
Dishman, Samuel, 83
Dixon, Jerema Sell, 39
Drury College, 34
Dryden, James, 25, 26
Duff, Bertha, 117
Duncan, James A., 53
Duncan, Mary, 53
Durham, Sarah (McQuerter), 55

E

Eslinger, Adam, 51
Evans, Alexander, 16, 27, 31
Evans, Daniel McCord, 28, 39
Evans, Eliza Elizabeth, 28, 31
Evans, Elizabeth Leathers, 28
Evans, John, 27, 31
Evans, Joseph, 15, 27, 28, 31
Evans, Melinda, 27

Evans Station, 22-23, 27
    location, 29-31
    photos, 29

F
Fair Play, Mo., 31-32, 33
Faul, George, 97
Fayette *Intelligencer*, 2
Fayetteville Road, 15
Fort Gibson, 3
Foster, John, 73-74
Foster, Samuel, 73
Fox tribe, 1, 2
Franklin, Mo., 4
Franklin *Intelligencer*, 2
Frankum, John, 117
Fulbright, William, 31, 48, 49, 51, 80

G
Gault, Mark, 119
Gay, Joshua, 100
Gay, William, 100
"Genealogies of Early Springfield Families", 64
Gibson, Josiah, 95
Goddard, Frederick B., 47-48
Gott, Joseph, 65, 73
Gott, Reese, 73
Greene County
    Confederate occupancy, 74
    County Court legislates roads, 15-16
    Section 14, 22-25
    Sections 35 and 36, 48-49
    Section 26, 63
    settlers moving, 47
    surveying of, 9, 49-50
Greene County Historical Society, 35
    mail route markers, 34-36
"Greene County Mills", 49

Greene County Missouri Deed Book, 66
Greene County Probate Book A, 116
Gurley, Martha, 85

H
Hagen, James, 79-80
Hagen, Tabitha Banfield, 79-80
Halbert, Jesse, 64, 65
Hall, Joseph, 78
Harmony Mission, 3
Hastings, W.T., 118
Hay, Henry, 16
Headlee, Elisha, 26
Headlee, John, 25
Headlee, Joseph, 71
Headlee, Nancy, 27
Headlee, Sam, 39
Headlee, Samuel G., 27
Highway 13, 30-31, 36, 42
History & Families, Polk County, Missouri, 104
History of Benton County, 5, 7
History of Cooper County, 38
History of Greene County, 34
History of Pettus County, Missouri, 38
History of the Greene County Alms House, 69
Hoffman, Mrs. (Murray family), 84
Hubble Personal Reminiscences, 34
Hurt, R. Douglas, 13

J
James, Thomas, 115
Jessup, Eli, 57
J.H. Creighton and Sons, 90
Johnson, Benjamin, 25, 27
Johnson, Sally Sims, 25
Jones, Henry G., 114
Jones, Lucy, 114
Jones, Madison, 114
Jones, William G., 61, 62, 110, 114

K

Kain, Joseph, 56
Kain, Letticia, 56
Kain, William L., 56
Kear, George W., 92
Kickapoo tribe, 2
Kidd, Polley, 78
Kilpatrick, Ann, 85
Kilpatrick (Kirkpatrick), Dorcus J., 85
Kilpatrick, Martha, 85
Kingman, W.A., 35

L

Lakeview Lighthouse Church, 72
Lanham, Amanda, 80
Lastley, Charity Johnson, 27, 49, 50
Lastley, Martha, 50
Lastley, Samuel, 14, 15, 24, 25, 27, 48, 49, 56, 74
    death, 50-51
"Lastley's field", 49-50
Lawrence County Historical Society Bulletin, 13
Lay, James H., 7
Leathers, Edward, 28
Leathers, Eliza, 28
Leathers, John, 28
Leathers, Nancy, 28
Leathers, Sarah, 28
Leathers, Theophilus, 25, 27, 28
Lee, Samuel, 26
Levens, Henry C., 38
Little Dry Sac River, 40
    bridge at ford, 40
Lloyd's Military Map, 33
Lost Hill Park, 83, 86
Louisiana Purchase, 1

M

MacDonald, Demuth I., 38
Makley, Eveline Murray, 103-104
Makley, John, 103
Makley, Marie, 103
Maple Park Cemetery, 53
Map of the Butterfield Overland Mail through Missouri, 22
Matt, Don, 21, 29
McCurdy, David, 71
McCurdy, Thomas, 73
McDonald, J.M., 95-96
McGrew, Roger, 85
McLaughlin, S.W., 98
McMurray, James, 68
McQuerter, Grace, 53
McQuerter, James Simerall, 48, 52-55, 81
McQuerter, Jane, 54, 55
McQuerter, J.M., 103
McQuerter, John, 54
McQuerter, Louisa, 54
McQuerter, Lucinda, 54
McQuerter, Malvina, 53
McQuerter, Nancy, 54
McQuerter, P.M., 54
McQuerter, Wesley, 54
McQuerter, William G., 53, 54
McQuirter, Laura, 53
Miller, Joseph H., 49
Mississippi River crossing, 5
Missouri Historical Society, 34-35
Mooney, John, 9
Morrison, Henry Clinton, 25, 26
Morrison, Nancy Sims, 25, 26
Mullings, Hosea, 26, 28, 31, 66, 68, 85
Mullins, George Cathey, 28, 31
Murray, Andrew J., 39, 73, 84, 86, 96, 97, 101, 102
Murray, Carl, 102
Murray, David C., 30, 39, 40, 55, 62, 66, 83-84, 89-90
    death, 58-59, 97
    farm, 105
    friendship with Tapley Daniel, 93-94

gravestone, 106
map, 23-24
move to Missouri, 90-93
Ohio home, 93
will, 96-97
Murray, Dorsey, 101
Murray, Edna, 102
Murray, Evalina (Eveline), 84, 96, 97, 101, 103-104 (daughter of David)
Murray, Evalina, 101, (daughter of Z.G.)
Murray, Ezenith, 39, 92, 96
Murray, Freddie, 108
Murray, Huldah, 39, 59, 92, 96, 104, 114. See Banfield, Huldah
death, 84
gravestone, 106
house, 123
Murray, Isaiah, 97, 104
Murray, Jasper, 39, 84, 96, 97, 100-101
Murray, Jennie, 112
Murray, Jettie, 102
Murray, Lena Brune, 101, 112
Murray, Lucy Jane Carter, 102
Murray, Luther, 101
Murray, Maggie Gay 100 (possible wife of Jasper)
Murray, Maggie Rosenberger, 101 (wife of Z.G.)
Murray, Maisie, 103
Murray, Mamie, 101
Murray, Margaret, 107, 108
Murray, Norman, 103
Murray, Ollie, 102
Murray, Ralph, 103, 112
Murray, Sarah Jane (Jenny?) Stivers, 102-103
Murray, Susan, 101
Murray, Susanna(h) Long, 92, 96, 104
Murray, Walter, 101
Murray, William, 39, 73, 84, 96, 97, 101, 112

Murray, William Penn, 102-103
Murray, Zelotus G. (Lotz), 39, 40, 84, 96, 97, 101-102, 106, 114
Murray, Z.G. 91, 92, 95 (brother of David), 90
Murray family cemetery
cleaners for, 110
current condition, 106-107
gravestones, 110
stones and documentation, 107-120

N

Nathan Boone and the American Frontier, 13
New York Herald, 20
New York Times, 34
New York Tribune, 90

O

Old Salem Cemetery, 52
Ormsby, J. Waterman, 20
Osage Indians
mission school for, 3
treaties, 2-3
Osage River
ferry, 5
steamboat traffic on, 8
Osage treaty of 1825, 2-3
Owen, Charles Baker, 80
Owen, J.S., 86
Owen, Sol H., CCJ, 79
Owen, Solomon, 80, 100
Owen Farm, 122
Ozarks Genealogical Society, 8, 19, 53, 72, 76, 106

P

Painter, Jacob, 115
Palmyra, Mo., 5
Panels, Elizabeth, 27
Past and Present of Greene County, Missouri, 7

Patterson, John A., 94
Phelps, John S., 35, 94
Pomme de Terre River, 5
post roads, Congress establishes, 8
Potawatomi tribe, 2
Potter, Benjamin, 83
Potter, Lucy Ann Banfield, 83, 85

R
Rabe, George B., 66
railroad, growth of, 38–39
Rector, Calvin, 73
Richardson, Albert D., 20–21
Rising, Marsha, 25, 52
Robberson, Allen, 28
Robberson, Bennett, 25
Robberson, Elizabeth Pettigrew, 25, 28
Robberson, Jane Caroline, 28
Robberson, Mary Evelyn, 28
Robberson Prairie, 28
Roberts, Lee, 73
Robertson (Robberson), William, 26
Robinson, B.H., 39
Robinson (Robberson), Rufus, 16
Rosenberger, Jacob, 101
Ross, James, 39
Rountree, Joseph, 115

S
Sac River bridges, 40, 42
Sac tribe, 2
Santa Fe Trail, 4
Saturday Morning Visitor, 8
Sauk tribe, 1
Sedalia, Mo., 38–39
Sell, Azenath A., 97, 104
Sell, Rolan, 104
Sims, Briggs, 25, 26
Sims, Burwell, 25
Sims, Eliza A. (Adams), 25, 27
Sims, Fanny S., 25, 16

Sims, John, 25
Sims, Mary Ann, 26
Sims, Robert, 25, 26
Sims, Washington, 25
Sims, Zachariah, 25, 27
Sloans, Jerimiah N., 9
Smith, Manda Ann (Amanda), 118, 119
Smith, Taylor, 119
South Pacific Railroad, 65, 66
Southern Pacific Railway, 32
Spain, Albert, 67
Spain, Anna E., 67
Spain, Claude, 67
Spain, Elizabeth Ann, 67
Spain, George, 67
Spain, James C. (Rev.), 62, 67
Spain, Jesse, 67
Spain, John, 67
Spain, Mary J., 67
Spain, Robert S., 67
Spain, William W., 67
Springfield, Mo.
    Butterfield Overland Mail route, 34
    immigration to, 91
    land office, 24
Springfield Daily News, 113
Springfield Leader Democrat, 68, 112, 113
Springfield Northwest Wastewater
    Treatment Plant, 52
Springfield Republican, 113
Springfield Weekly Patriot, 90, 91, 117
St. Louis - Boonville post road, 3–4
Stark, Charles, 57
Sterling, Ethel, 104
Sterling, George, 104
Sterling, Hamilton, 104
Sterling, Jerema Murray, 97, 104

T
Terrell, Rachel Banfield, 71, 79, 85
Terrell, Thomas, 71, 79

24th Missouri Volunteer Infantry, history of, 74
The Pictorial and Genealogical Record of Greene County, 31
The 24th Missouri Volunteer Infantry, 61
The Western Tourist or Emigrant's Guide, Mississippi Valley, 47
Thornhill, Charles, 73-74
Thornhill, Epperson, 58
Thornhill, Hiram, 57-58
Treaty of 1808, 1-2

U
United Foreign Missionary Society, 3

V
Vaughan, Cora, 118, 119
Vaughan, Edward, 119
Vaughan, Eliza, 119
Vaughan, Maria, 118, 119
Vaughan, Tyler, 119

W
Wallis, Jeptha, 52
Walters, Charley, 86
Warren, Ann, 62, 82
Warren, William, 78
Washam, Willis, 81
Wells Fargo, 38
West, J. Dale, 13
West, movement to, 90-91
Where to Emigrate and Why, 47-48
White, A.D., 57
White, Caledonia, 60
White, Elizabeth Jane, 66
White, H.C. (Hardy?), 97
White, Hugh L., 60
White, James, 73
White, James A., 60
    (son of Joseph B. White)
White, Joseph Bolivar, 52, 57, 60
White, Julia E., 60
White, Martha Daniel, 57, 60
White, Sophronia, 60
White, Walter, 60
White, William, 81
Williams, Anne Banfield, 79
Williams, G., 60
Williams, Lemuel, 79
Williams, Lucinda, 60
Williams, Victoria, 60
Williamson, C.C., 62
Wilson, Margaret Bond, 26
Wilson, Mary, 28
Wilson, Thomas, C. 25, 26, 28
Wilson, T.S., 39, 84-85, 86
Wilson, William, 52
Wrought Iron Bridge Company, 40, 42
Wyandot Democratic Union, 89

www.ingramcontent.com/pod-product-compliance
Lightning Source LLC
Chambersburg PA
CBHW050643160426
43194CB00010B/1791